DESIGN & MANAGEMENT For RESOURCE RECOVERY

Volume 3
QUANTITATIVE DECISION-MAKING

DESIGN & MANAGEMENT For RESOURCE RECOVERY

Volume 3 QUANTITATIVE DECISION-MAKING

by

Albert J. Klee

P. Aarne Vesilind
Series Editor

ANN ARBOR SCIENCE
PUBLISHERS INC / THE BUTTERWORTH GROUP

230 Collingwood, P.O. Box 1425, Ann Arbor, Michigan 48106

Library of Congress Catalog Number 80-67973
ISBN 0-250-40313-7

Manufactured in the United States of America

SERIES PREFACE

Given today's economic, social and political indicators, pessimism seems to be the path of least resistance. The United States and the rest of the western world are already importing most of the requisite raw materials to feed the industrial state. Nations that own these resources, recognizing the economics of scarcity, will charge ever-increasing prices for them, thus fueling worldwide inflation. In a material- and energy-scarce society, Keynesian economics no longer apply, and inflation, unemployment and a steadily decreasing standard of living seems to be part of our future.

Much of this pessimism, however, is based on the assumption that the resource use cycle is one-way and open-ended—materials are extracted, used and discarded, to be lost forever. Fortunately, this is not necessarily a correct assumption, for it is possible to reclaim useful materials from waste and inject these back into the resource cycle. Likewise, much of the organic fraction in wastes, which derives its energy from the inexhaustible energy source—the sun—can be burned to produce useful heat or processed to yield other fuels.

The science and technology associated with such a reclamation of materials and energy from waste has become known as resource recovery. It is a fledging technology, borrowing unabashedly from other fields such as mining engineering, biochemistry, chemical process engineering and thermal sciences. Although it has experienced growing pains (often as the result of overenthusiasm and inadequate technical and managerial skills), it nevertheless seems to be an absolute requisite for an orderly transition through the 1980s and into the 21st century.

The series *Design and Management in Resource Recovery* is the first continuing series of books devoted to this topic. The authors all have wide-ranging knowledge in their individual areas of expertise and have produced volumes which are both contemporary and useful as reference works for years to come.

Reading and editing the series is not only a rewarding experience, but has substantially abated my feelings of pessimism about the future. Learning to design and manage resource recovery systems will not guarantee us a happy and prosperous future, but it is at least one meaningful bridge that will help bring it about.

P. Aarne Vesilind

PREFACE

This book examines a wide variety of innovative approaches to some rather difficult resource recovery decision problems. That these approaches are unusual in the literature of resource recovery is evident from a glance at the references listed at the end of each chapter. In a book with the word "quantitative" in its title, one expects to find citations dealing with engineering, statistical and mathematical journals, but the lists also include such publications as *Socio-Economic Planning Sciences, Journal of Mathematical Psychology, Management Science, Harvard Business Review, Technological Forecasting & Social Change* and *Behavioral Science*, to name just a few.

Because the scope of this book is clearly multidisciplinary in nature, it was realized that this could pose a problem for the reader who was unfamiliar with these areas. For this reason, I have taken a pragmatic, applied approach to the material, using many numerical examples to illustrate the suggested techniques. Although an attempt has been made to provide a somewhat heuristic development of the theory behind each of these techniques, it was neither possible nor advisable to expound on theory in any great detail. I state the Propagation of Error formula in Chapter 1, for example—a technique not likely to be known to many engineers—but do not provide a proof because the three extra pages of mathematical development required would only delay the reader faced with the task of solving immediate real-world problems. References are provided at all times, however, for those who are interested and have the time to review the background theory in greater depth.

There are a number of detailed proofs in the appendixes to certain chapters that, with regard to mathematical manipulation, admittedly do not spare the reader. However, these are only included because they provide theoretical support to original work specifically developed for this text by the author, or because they are more complete than those available elsewhere. Other than this, the reader with a basic knowledge of the concepts of probability and statistics should have no real difficulty in understanding and applying the methodologies discussed in this book.

Albert J. Klee

ACKNOWLEDGMENTS

The mathematical and statistical research that supports the quantitative developments of this book have been developing over many years, and most of what a man knows or writes is derived from other men's hard work. My great and overwhelming debt is to these contributors to the literature in probability, statistics, decision theory and mathematics, as well as to those investigators who have applied these sciences to practical problems. I acknowledge their contributions collectively rather than individually.

I would like to express my personal appreciation to my friends and colleagues of Committee E-38 on Resource Recovery, American Society for Testing and Materials, for first bringing to my attention the need for new approaches to the sampling of solid wastes for quantity and composition, and for their encouragement of my research in this area. Nor should the efforts be overlooked of Dr. P. Aarne Vesilind, of Duke University, who provided me the opportunity to mix the many ordinarily immiscible disciplines involved in solid waste decision-making. My wife, Harriet, deserves my greatest thanks for her seemingly inexhaustible supply of love and encouragement during the writing of this book.

Finally, the opinions expressed herein are solely those of the author and should not be construed as representing in any way the policies of the U.S. Environmental Protection Agency. Although there are undoubtedly errors of both omission and commission, it is hoped that they are not too serious. I, of course, accept responsibility for them all.

Albert J. Klee is a Senior Environmentalist in the U.S. Environmental Protection Agency Office of Research and Development, and an Adjunct Associate Professor of Management at Xavier University, Cincinnati.

Dr. Klee received a PhD (interdisciplinary in Systems Analysis, Economics and Public Administration) from the University of Cincinnati; an MChE from New York University; an MBA and MS in Mathematics from Xavier University; and a BChE from City College, New York.

Dr. Klee's primary research interests have focused on resource recovery, hazardous wastes and the application of decision theory, statistics and systems analysis to government operations and environmental problems. He has written numerous technical papers, which have appeared in professional journals, books and conference proceedings. Professional memberships include the American Statistical Association, Institute of Management Sciences, American Society of Ichthyologists and Herpetologists, American Institute of Chemical Engineers, and Air Pollution Control Association.

Dr. Klee is also a part-time explorer and amateur ichthyologist. He has explored the jungles and savannas of Africa, South America and Central America, and has discovered two new genera of fishes and several new species. A species of fish, *Apistogramma kleei*, found in the Amazon Basin, is named after him.

CONTENTS

CHAPTER 1

ESTIMATION OF RAW REFUSE QUANTITY

1.1 INTRODUCTION

This book is concerned primarily with innovative quantitative approaches to resource recovery decisions, so it might seem unusual that the first chapter deals with what appears to be a rather traditional subject, i.e., that of sampling to determine the quantity of solid waste available in a given waste shed over a stated period of time. (In this context, a waste shed is the area from which waste is expected to be available for a proposed resource recovery system.)

There are already many excellent texts available that deal with survey sampling [1,2]; however, the sampling techniques described in these texts do not fit the circumstances when the problem is to estimate the quantity of raw waste arriving at a given site, such as a landfill, transfer station or incinerator pit. For one thing, the sample comes to the investigator, which is the reverse of the situation commonly described in standard textbooks. Further, sample size varies in quantum jumps in the solid waste case because of the measurement systems that must be used. We cannot rent one-half of a wheel or platform scale; it is either the whole scale or nothing. Once a scale is rented and the labor hired to make the measurements, it makes good statistical sense (as we shall see) to use the equipment and labor to the fullest extent possible. This is totally unlike survey sampling where the variable cost of an interview is usually greater than the fixed cost. Moreover, solid waste generation exhibits a strong seasonal effect, and such effects must be taken into consideration in our sampling schemes. A discussion of techniques for dealing with this problem, however, is noticeably absent in the literature of resource recovery.

Finally, for a number of sound reasons, traditional survey sampling usually attempts to select unbiased samples. It will be shown, however, that where solid waste is involved it is more efficient to select biased samples, then correct for the bias in the estimation formulas employed. Thus, it is appropriate to take a fresh look at the problem of estimating quantity of raw solid waste for resource recovery purposes.

It might be well at this point to comment on alternative paths to the estimation of solid waste quantity. There are two basic approaches: (1) direct measurement, and (2) predictive models. Within both approaches are many variations on a theme. In direct measurement, for example, one can measure at the point of generation (i.e., at each house, commercial establishment, etc.) or at the point of disposal (landfill, incinerator, community recycling center, etc.). The former is obviously much more costly and time consuming. Indeed, as the "interviewer" literally goes to the sample, point of generation methods more closely resemble traditional survey sampling. Thus, the selection of the sample is more complex than in point of disposal methods. Further, since our objective here is to estimate the quantity of solid waste that would be available for a resource recovery system, it would be necessary to determine its destination as well. In other words, just because a bundle of newspapers is set out in front of a residence or a stack of wooden crates placed behind a commercial establishment does not necessarily mean that they are candidates for a possible resource recovery system. Such waste might be destined for recycling or as a raw material for some process, and might not be available for other purposes, such as the resource recovery system the designer has in mind.

Predictive models rely on surrogate or ancillary measurements to estimate waste quantity. The most complicated are the input-output models, which track the goods entering a waste shed and determine the conversation rate and quantity of waste produced from them. Such approaches are infeasible here because (1) the data bases simply are not available, and (2) it would require enormous resources in computer facilities to implement. The least complicated predictive models are those in the form of predictive equations, in which the only inputs are the values of the surrogate variables. The simplest example is one based on per capita figures. If the waste generation is 5 lb/person/day, for example, all that is needed to estimate quantity is a knowledge of the population in the waste shed. Like the point of generation version of direct measurement, however, such models do not tell us whether the waste is available. Further, there are many other variables that affect waste generation; geography, climate, income level, local ordinances, etc. To obtain models with any statistical validity, much data would have to be gathered all over the country to estimate the parameters (constants) of the model. Also, the significance of the individual variables and the values of the parameters would be expected to change with time. Therefore, such models would have to be maintained, a costly procedure that has little chance for support even by government agencies. The author's conclusion, therefore, is that the most pragmatic approach to the estimation of waste quantity for resource recovery purposes is direct measurement at the point of ultimate disposal, such as the landfill.

1.2. PRELIMINARY STATISTICAL CONSIDERATIONS

In statistical sampling, the elements of a population are defined jointly with the population itself, i.e., the elements are the basic units that comprise and define the population [1]. When sampling at sites such as a landfill, transfer station or incinerator pit, the population elements usually can be taken as the individual vehicle-loads. It is important not to confuse the vehicles with the vehicle-loads. One might observe 500 loads delivered to a given landfill in one week, but involving only 100 different vehicles.

A population value or a "parameter" is a numerical quantity that summarizes the values of some characteristic for all elements of the population. Generally, our interest is in the total quantity of solid waste delivered to the site over a given period of time (generally a year), although, as we shall see, there is interest also in the average load per vehicle. Precision and accuracy are two terms that are relevant to the parameters of a population; more specifically, they are terms applied to *estimates* of the parameters. An estimate is accurate if it is close to the true value; the difference between the estimate and the true value is known as the "bias." Precision, on the other hand, refers to repeated measurements, i.e., more than one estimate of the same parameter. If the estimates do not vary much among themselves, the method of estimation is said to be precise. The concept of precision is inversely related to that of variance or standard deviation. In other words, the greater the precision, the less the variance or standard deviation of the estimate. Precision is related to the uncertainty of an estimate. These concepts are illustrated, in a relative sense, by the distributions shown in Figure 1-1. Each of the figures represents the distribution of repeated estimates of some true value, T, and we assume that the estimate of T is taken as the mean of the distributions shown. It will be noted that the accurate distributions (a and c) are centered on the true value, T, i.e., the bias is zero. The precise distributions (a and b) are those with less dispersion, i.e., smaller standard deviation. Obviously we would prefer estimates that are both unbiased and of high precision.

Precision and accuracy are affected by two things: (1) the selection process, i.e., the manner in which we select our sample; and (2) the estimation process, i.e., the method by which we compute our estimates from our sample. However, the major problem in determining raw solid waste quantity involves the selection process.

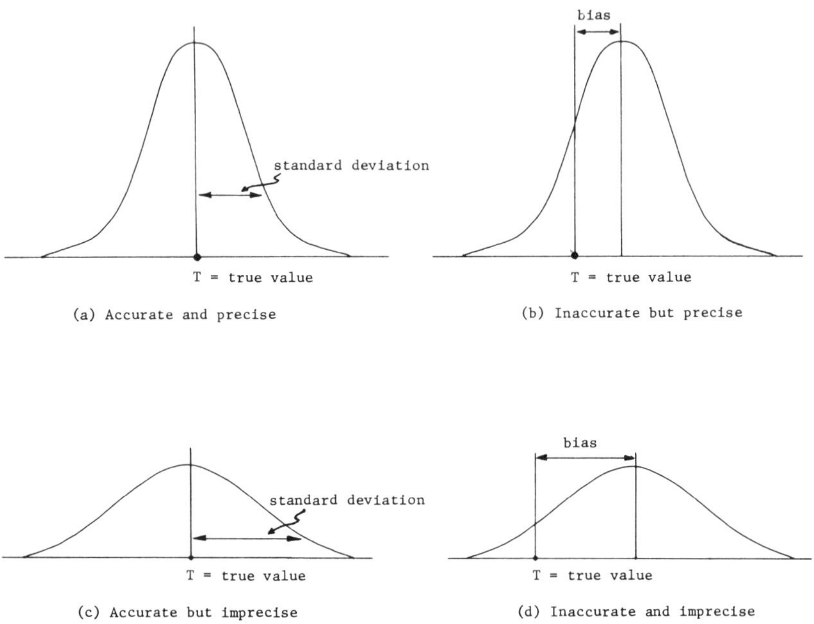

Figure 1-1. Concepts of accuracy and precision.

1.3. THE SELECTION PROCESS

For the moment, we will defer both the problems of trend and cyclical or seasonal variation of the quantity of raw solid waste delivered to a given site and assume that sampling is to take place over a period of one week. By sampling for a one-week period, variations in quantity due to hour-to-hour and day-to-day differences are accounted for in our estimate. Because week-to-week differences are also very significant, it makes little sense to sample for periods of two or more *consecutive* weeks. Simply stated, if we can afford (in both the economic and the time sense) to sample for more than one week, it pays to sample for one-week periods spread throughout the year. This is treated in greater detail in Sections 1.9 and 1.10. In any event, the problem of bias in the selection process can be addressed generally in either of two ways:

1. Make no assumptions about the nature of the arrivals of the vehicle loads and randomize the sample. Or,
2. Assume that the vehicles arrive in random order and take a systematic sample, which is one in which every kth vehicle arriving at the site is weighed.

The second approach is attractive for several reasons: (1) the procedure by which the trucks are selected for the sample is relatively simple and "foolproof," and (2) if we are interested in separate estimates for different types of vehicle-loads (i.e., different sizes of trucks, commercial versus residential, Firm A versus Firm B, etc.), the systematic sample can easily yield a proportionate sample, which has statistical advantages that will be explained later.

The basic problem with systematic sampling has to do with the departure from the assumed randomness in the arrival of the vehicles. These departures are of two kinds:

1. A monotonic trend may exist in the weights of the loads. For example, the loads may increase with time over the sampling period. Since a systematic sample consists of a random start followed by sampling each truck at interval k afterwards, the estimate will depend on the random start within the first interval. In Figure 1-2a, for example, the low random start (solid dots) will produce a lower estimate than the high random start (open circles). The estimates in these two cases will be biased either low (solid dots) or high (open circles).
2. A cyclical or periodic trend may exist in the weights of the loads (Figure 1-2b). In Figure 1-2b, if the random start happens to fall at the top of a cycle (solid dots), the estimate will be high; if it falls at the bottom of a cycle (open circles), it will be low. Again, in either case the estimates will be biased.

It does not appear, however, that either of these events is particularly likely. If a monotonic trend did exist, the sampling period of one week is so short that the effect and the bias would be negligible. Periodic fluctuations are potentially more harmful, but it is also highly unlikely that the frequency would be constant over a period of one week if one did exist. If either condition *did* exist, we could simply change the random start several times during the week, as this would reduce the effect of any single start.

1.4 SYSTEMATIC AND RANDOM SAMPLING

Systematic sampling simply consists of taking every kth sampling unit (vehicle-load) after a random start. The random start, r, is the rth vehicle load in the arriving vehicle sequence where r is a number, chosen at random, between 1 and k. The succeeding vehicles to be sampled are $k + r$, $2k + r$, $3k + r$, etc. The random start from 1 to k imparts to each vehicle-load the selection probability of $1/k = f$ where f is known as the sampling frequency. If we know the total number of vehicle-loads, N, arriving during the sampling period, the total sample size, n, is given as $n = fN = N/k$. (It will be noted that n will be an integer only if N is an integral multiple of k. See the last part of Section 1.9.)

Although the concept of systematic sampling is relatively simple to implement in theory, a problem arises when sampling is scale-limited. For example,

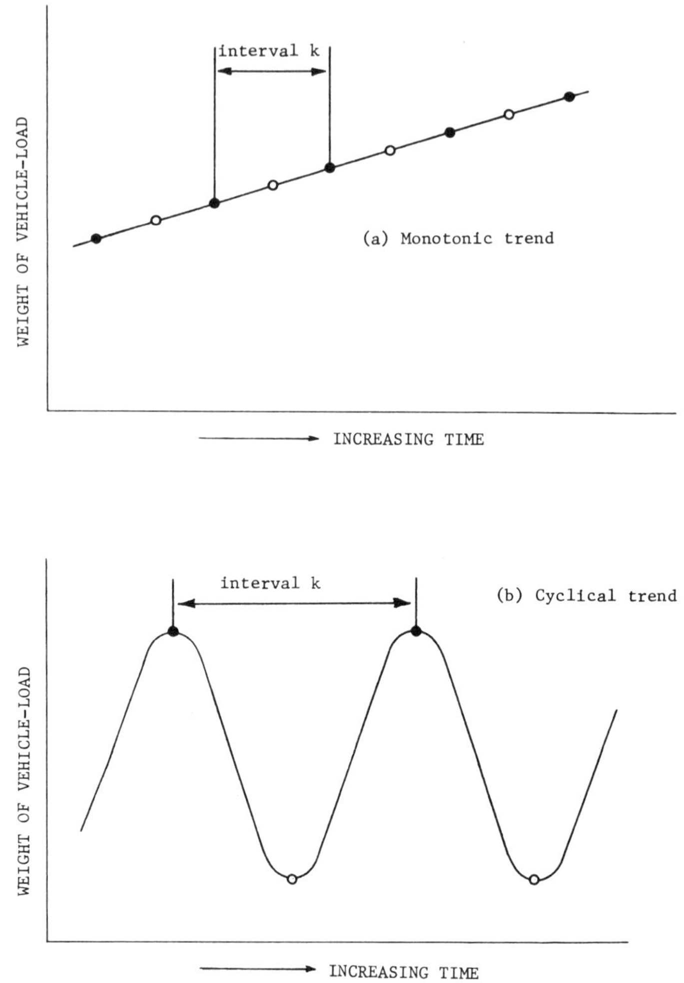

Figure 1-2. Effects of nonrandomness on systematic samples.

suppose it is known that approximately 750 vehicle-loads arrive at a given landfill over a five-day period and a sample size of 75 is desired. Since k = N/n, the sampling interval is every 10 vehicles (after a random start of 1–10). Suppose further that it is practicable to weigh up to 10 trucks per hour without seriously interfering with vehicular flow. It would appear, then, that no problems would occur because the average interarrival time of vehicles is (5 days) (8 hr/day) (60 min/hr)/(750) = 3.2 minutes. On the average, then, one vehicle would be weighed every 32 minutes, well within the capability of the scale system. Unfortunately, vehicles do not have uniformly distributed interarrival times, i.e., there may be peak arrival periods when the number of trucks arriving and to be sampled overwhelms the weighing capability. We are forced to "default" on weighing some of the trucks selected by the sampling plan. This has two effects. One result is that fewer trucks are weighed than our sampling plan calls for; clearly this reduces the precision of our estimate. Second, if the load weights of the defaulted samples differ appreciably from the nondefaulted samples, bias will be introduced. For example, at many landfills, trucks arriving toward the end of the day tend to have lower load weights than those arriving at other times. (This is simply a matter of a policy not to have vehicles stand overnight with refuse in them.) Since fewer vehicles arrive toward the end of the day, the tendency would be to oversample the lightly loaded vehicles and to undersample the normally loaded vehicles. Thus a bias is introduced.

The selection of a random sample is more complicated than that of a systematic sample. Assuming a selection probability equal to that of a systematic sample, we must sample at random with probability $f = 1/k$. If we desire a 10% sample (i.e., $f = 0.10$), then we must consider *each* truck entering the site and, using a probability generator of some sort, decide whether to sample it. (One could use a table of random numbers from 1-1000, for example. If the random number fell below 101, we would sample the truck; otherwise we would let it pass.) Not only is this more complicated than systematic sampling, it may result in a greater number of defaulted samples due to the "bunching up" effect of the random sample as opposed to the systematic sample. However, the effect on accuracy and precision due to defaulting in a random sample is exactly the same as in a systematic sample.

1.5 CORRECTING FOR DEFAULTING

For a simple random sample, or for a systematic sample where it can be assumed that the population contains neither trend or cyclical components, the mean, $\bar{X}$, is estimated by

$$\bar{X} = \sum^{n} w_i X_i = 1/n \sum^{n} X_i \qquad (1\text{-}1)$$

where X_i is the ith observation (i.e., weight of the ith vehicle-load) and the weight, w_i is equal to $1/n$, where n is the total number of observations. The total, X, is estimated by

$$X = N\bar{X} \qquad (1\text{-}2)$$

where N is the total number of vehicle-loads in the population for the time sampled (we will assume that the sampling time is one week). The variance of the individual loads is computed as follows:

$$\mathrm{var}(X_i) = \frac{\sum^{n} X_i^2 - \left(\sum^{n} X_i\right)^2 \Big/ n}{n-1} \tag{1-3}$$

and the variance of the mean and of the total are

$$\mathrm{var}(\bar{X}) = \frac{1-f}{n}\,\mathrm{var}(X_{ij}) \tag{1-4}$$

$$\mathrm{var}(X) = N^2 \mathrm{var}(\bar{X}) \tag{1-5}$$

where $f = n/N$. We can restate the formula for the estimation of a mean by grouping the observations on an hourly basis, i.e.,

$$\bar{X} = \sum_{i}^{h} \sum_{j}^{n_i} w_{ij} X_{ij} = 1/n \sum_{i}^{h} \sum_{j}^{n_i} X_{ij} \tag{1-6}$$

where X_{ij} is the jth observation in the ith hour, n_i is the number of observations in the ith interval, h is the number of hours, n = the sum of the h n_i, and $w_{ij} = 1/n$. The estimate of the variance of an individual truck load now becomes,

$$\mathrm{var}(X_{ij}) = \frac{\sum_{i}^{h} \sum_{j}^{n} X_{ij}^2 - \left(\sum_{i}^{h} \sum_{j}^{n} X_{ij}\right)^2 \Big/ n}{n-1} \tag{1-7}$$

Equations 1-2 and 1-5 remain unchanged.

Now consider the case where we undersample during the hourly time periods. This would occur if sampling were limited by scale capacity. If there is lack of randomness in the data (as there would be, for example, if the loads tended to weigh less toward the end of the day), the equal weighting of observations implied in Equation 1-6 would bias the estimate of the mean. We can, however, modify the weight of each observation by the quantity $m_i/(n_i m)$, where m_i is the total number of trucks arriving in the ith hour, and m is the total number of trucks arriving during all the hours sampled. Clearly this removes the bias when $m_i \neq n_i$. Introducing these weights into Equation 1-6 we have

$$\bar{X} = \sum_{i}^{h} \sum_{j}^{n_i} m_i X_{ij}/n_i m = \sum_{i}^{h} \sum_{j}^{n_i} w_i' X_{ij} \tag{1-8}$$

where $w_i' = m_i/n_i m$ and m is the sum of the h m_i. Note that if n_i/m_i is some constant, k (i.e., proportionate sampling, which is unbiased), then w_i' =

$1/km = i/n = w_i$. A proof that $\overline{X}$ in Equation 1-8 is an unbiased estimate of the true mean is given in the appendix (A.1) to this chapter.

To find the variance of $\overline{X}$ in Equation 1-8, we utilize the concept of propagation of error [2], which states the following:

$$\text{var}[f(x_i, x_2, \ldots, x_n)] = \sum_{i}^{n} \left(\frac{\partial f}{\partial x_i}\right)^2 \text{var}(x_i) + \sum_{i}^{n} \sum_{\substack{j \\ i \neq j}}^{n} \frac{\partial f}{\partial x_i} \frac{\partial f}{\partial x_j} \text{cov}(x_i, x_j) \tag{1-9}$$

Applying the propagation of error formula to Equation 1-8 we obtain

$$\text{var}(\overline{X}) = \frac{1}{m^2} \text{var}(X_{ij}) \sum_{i}^{h} m_i^2 (1 - f_i)/n_i \tag{1-10}$$

where $f_i = n_i/m_i$. The rather complicated derivation of Equation 1-10 is given in the appendix (A.2) also. Equations 1-2 and 1-5 remain unchanged.

The use of these formulas is illustrated in Table 1-1, where data from a random or a systematic sample are presented for an 8-hour sampling period.

Table 1-1. Calculations for Example of Section 1.5

Hour	(1) n_i	(2) ΣX_{ij}	(3) m_i	(4) w_i'	(5) $\Sigma w_i' X_{ij}$[a]	(6) $m_i^2 (1 - f_i)/n_i$
1	1	12,452	10	0.0333	414.6516	90
2	1	10,223	10	0.0333	340.4259	90
3	2	37,260	20	0.0333	1,240.7580	180
4	6	83,613	80	0.0444	3,712.4172	986.67
5	6	101,431	100	0.0556	5,639.5636	1566.67
6	3	56,305	30	0.0333	1,874.9565	270
7	3	24,418	30	0.0333	813.1194	270
8	2	26,181	20	0.0333	871.8273	180
Totals	24	351,883	300		14,907.7195	3633.34

We assume that the standard deviation of these 24 samples is 5408 and that N = 300

Case I: Random or Systematic Sample, No Scale Constraints

$$\overline{X} = 351{,}883/24 = 14662 \text{ using equation 1-6}$$

$$\text{Standard deviation (SD) } (\overline{X}) = \sqrt{1 - (24/300)}\ 5408/\sqrt{24} = 1059 \text{ using equation 1-4}$$

Case II: Random or Systematic Sample, Scale Constraints

$$\overline{X} = 14{,}907.7 \text{ using equation 1-8}$$

$$\text{SD } (\overline{X}) = \frac{\sqrt{3633.34}}{300} (5408) = 1087 \text{ using equation 1-10}$$

[a]The weight term X_{ij} is expressed here in pounds. The adoption of another mass term would not change the example, and the use of dual (SI and common) units has been avoided to facilitate clarity and enhance the usefulness of the example.

The standard deviation of 5408 is the standard deviation of all 24 observations, as it is not necessary to calculate this for each time period. Assuming that the n_i shown in the second column are equal to the target sample numbers (i.e., there is no scale restriction on weighing), Equation 1-6 is used to calculate the mean and Equation 1-4 to calculate the variance (the standard deviation, which is the square root of the variance, is shown in the table instead). If, because of scale limitations, we do not attain the target values, n_i, Equations 1-8 and 1-10 are used to calculate the mean and variance, respectively. $\bar{X}$ is greater in Case II than in Case I, mostly because a greater weight is assigned to hour 5, which has a mean of 16,905 i.e., higher than the average. Note that there is also an increase in standard deviation in Case II. This is due to the weighting factors used to unbias the estimate of the mean, and is the price we pay to remove the bais.

1.6 OVERSAMPLING

Typically, depending on the method of readout and the number of axles involved, it requires up to 30–45 minutes to weigh one truck using wheel scales. However, platform scales are 20 to 30 times as fast. When using platform scales, then, it will usually be possible to oversample some intervals, if not all of them. Two questions then arise: (1) If the intervals are oversampled, how are the estimates adjusted for bias? and (2) Is there any advantage to oversampling? The answer to the first question is simply that we use the same equations as for undersampling, i.e., Equations 1-8 and 1-10 (the equations for totals, i.e., 1-2 and 1-5, also apply). As for the second question, when we oversample, the variance *decreases*. (The proof of this is given in Appendix A.3.) Thus it always pays to oversample. (It is assumed that the scales are rented for a given time period and that no extra people need to be hired.) As an example, if in Table 1-1 all the n_i were increased to 6 for a total sample size of 48, and assuming the standard deviation of X_{ij} did not change, $\Sigma m_i^2(1 - f_i)/n_i$ would become 2900 and the standard deviation of $\bar{X}$ would be 971, a reduction of almost 11%.

1.7 COMPUTER SIMULATION MODEL

To illustrate the sampling plans described and to show the effects of the variation of critical parameters, a typical landfill operation was simulated using a General Purpose Simulation System (GPSS)-360 [3]. This program (Figure 1-3) can also be run on a GPSS-V processor. All the parameters and distributions used in this model are based on actual data [4], although it

```
*   REFUSE TRUCK SIMULATION
         SIMULATE
 WEIGH STORAGE    2              TWO WHEEL SCALES
 SLOW  FVARIABLE  (X7*X14)/100   MODIFICATION FOR END OF DAY
 TABL1 TABLE      X8,0,10000,3   TABLE FOR ALL TRUCKS ACCUMULATED
 TABL2 TABLE      X8,0,10000,3   TABLE FOR SAMPLED TRUCKS ACCUMULATED
 TABL3 TABLE      X8,0,10000,3   TABLE FOR ALL TRUCKS INCREMENTAL
 TABL4 TABLE      X8,0,10000,3   TABLE FOR SAMPLED TRUCKS INCREMENTAL
 1     FUNCTION   RN1,C6      TRUCK TYPE 1
0,3378/.2,7892/.4,9697/.6,11101/.8,12305/1,13408
 2     FUNCTION   RN2,C6      TRUCK TYPE 2
0,4316/.2,7775/.4,11234/.6,14692/.8,18151/1,21610
 3     FUNCTION   RN3,C6      TRUCK TYPE 3
0,4089/.2,12704/.4,16150/.6,18830/.8,21127/1,23232
 4     FUNCTION   RN4,C6      TRUCK TYPE 4
0,5594/.2,9563/.4,13532/.6,17501/.8,21470/1,25439
 5     FUNCTION   RN5,C6      TRUCK TYPE 5
0,0/.2,4054/.4,8110/.6,12163/.8,16218/1,20272
 6     FUNCTION   RN6,C6      TRUCK TYPE 6
0,7515/.2,11173/.4,14830/.6,18488/.8,22145/1,25803
 7     FUNCTION   RN1,C24     EXPONENTIAL DISTRIBUTION
0,0/.1,.104/.2,.222/.3,.355/.4,.509/.5,.69/.6,.915/.7,1.2/.75,1.38
.8,1.6/.84,1.83/.88,2.12/.9,2.3/.92,2.52/.94,2.81/.95,2.99/,96,3.2
 97,3.5/.98,3.9/.99,4.6;.995,5.3/.998,6.2/.999,7/.9998,8
 8     FUNCTION   RN7,D3         DETERMINE RESIDIENTIAL TRUCK TYPE
.07,1/.16,2/1,3
 9     FUNCTION   RN8,D3         DETERMINE COMMERCIAL TRUCK TYPE
.69,4/.85,5/1,6
       INITIAL    X9,4           RANDOM START
       INITIAL    X10,10         SAMPLING INTERVAL
       INITIAL    X11,563        STARTING MEAN FOR RES DIST
       INITIAL    X12,2250       MEAN FOR WEIGHING TIME
       INITIAL    X13,450        MODIFIER FOR WEIGHING TIME
       INITIAL    X14,100        FACTOR FOR END OF DAY
       INITIAL    X15,999999     STARTING MEAN FOR CUMM DIST
       RMULT      5,47,879,773,423,351,747,99
 ABLE  GENERATE   X11,FN7        GENERATE RESIDENTIAL TRUCKS
       ASSIGN     1,FN8          ASSIGN RESIDENTIAL TYPE
       TRANSFER   ,CONT
 BAKER GENERATE   X15,FN7        GENERATE COMMERCIAL TRUCK
       ASSIGN     1,FN9          ASSIGN COMMERCIAL, TYPE
 CONT  SAVEVALUE  7,FN*1         FIND TRUCK WEIGHT
       SAVEVALUE  30,V$SLOW      ADJUST WEIGHT FOR END OF DAY
       SAVEVALUE  29,P1
       PRINT      ,,C,D          PRINT CLOCK TIME
       PRINT      29,30,X,D      PRINT TRUCK TYPE AND WEIGHT
       TERMINATE
       GENERATE   3600           RUN FOR 1 HOUR
       TERMINATE  1
       START      1
 ABLE  GENERATE   X11,FN7
 BAKER GENERATE   X15,FN7
       INITIAL    X15,878
       RESET      TB$TABL1-TB$TABL2
       START      1
 ABLE  GENERATE   X11,FN7
 BAKER GENERATE   X15,FN7
       INITIAL    X11,281
       RESET      TB$TABL1-TB$TABL2
       START      1
 ABLE  GENERATE   X11,FN7
 BAKER GENERATE   X15,FN7
       INITIAL    X11,94
```

Figure 1-3. Within-week refuse truck simulation for computer model of Section 1.7. (This figure continues on the four following pages.)

FIGURE 1-3 (contd.)

```
      INITIAL   X15,439
      RESET     TB$TABL1-TB$TABL2
      START     1
ABLE  GENERATE  X11,FN7
BAKER GENERATE  X15,FN7
      INITIAL   X15,293
      RESET     TB$TABL1-TB$TABL2
      START     1
ABLE  GENERATE  X11,FN7
BAKER GENERATE  X15,FN7
      INITIAL   X11,281
      INITIAL   X15,125
      RESET     TB$TABL1-TB$TABL2
      START     1
ABLE  GENERATE  X11,FN7
BAKER GENERATE  X15,FN7
      INITIAL   X11,563
      INITIAL   X15,220
      INITIAL   X14,83
      RESET     TB$TABL1-TB$TABL2
      START     1
ABLE  GENERATE  X11,FN7
BAKER GENERATE  X15,FN7
      INITIAL   X15,439
      RESET     TB$TABL1-TB$TABL2
      START     1              END OF MONDAY
ABLE  GENERATE  X11,FN7
BAKER GENERATE  X15,FN7
      INITIAL   X11,655
      INITIAL   X15,999999
      INITIAL   X14,100
      RESET     TB$TABL1-TB$TABL2
      START     1
ABLE  GENERATE  X11,FN7
BAKER GENERATE  X15,FN7
      INITIAL   X15,1029
      RESET     TB$TABL1-TB$TABL2
      START     1
ABLE  GENERATE  X11,FN7
BAKER GENERATE  X15,FN7
      INITIAL   X11,327
      RESET     TB$TABL1-TB$TABL-2
      START     1
ABLE  GENERATE  X11,FN7
BAKER GENERATE  X15,FN7
      INITIAL   X11,109
      INITIAL   X15,514
      RESET     TB$TABL1-TB$TABL2
      START     1
ABLE  GENERATE  X11,FN7
BAKER GENERATE  X15,FN7
      INITIAL   X15,343
      RESET     TB$TABL1-TB$TABL2
      START     1
ABLE  GENERATE  X11,FN7
BAKER GENERATE  X15,FN7
      INITIAL   X11,327
      INITIAL   X15,147
      RESET     TB$TABL1-TB$TABL2
      START     1
ABLE  GENERATE  X11,FN7
BAKER GENERATE  X15,FN7
      INITIAL   X11,655
      INITIAL   X15,257
      INITIAL   X14,83
```

FIGURE 1-3 (contd.)

```
      RESET     TB$TABL1-TB$TABL2
      START     1
ABLE  GENERATE  X11,FN7
BAKER GENERATE  X15,FN7
      INITIAL   X15,514
      RESET     TB$TABL1-TB$TABL2
      START     1              END OF TUESDAY
ABLE  GENERATE  X11,FN7
BAKER GENERATE  X15,FN7
      INITIAL   X11,655
      INITIAL   X15,999999
      INITIAL   X14,100
      RESET     TB$TABL1-TB$TABL2
      START     1
ABLE  GENERATE  X11,FN7
BAKER GENERATE  X15,FN7
      INITIAL   X15,1029
      RESET     TB$TABL1-TB$TABL2
      START     1
ABLE  GENERATE  X11,FN7
BAKER GENERATE  X15,FN7
      INITIAL   X11,327
      RESET     TB$TABL1-TB$TABL2
      START     1
ABLE  GENERATE  X11,FN7
BAKER GENERATE  X15,FN7
      INITIAL   X11,109
      INITIAL   X15,514
      RESET     TB$TABL1-TB$TABL2
      START     1
ABLE  GENERATE  X11,FN7
BAKER GENERATE  X15,FN7
      INITIAL   X15,343
      RESET     TB$TABL1-TB$TABL2
      START     1
ABLE  GENERATE  X11,FN7
BAKER GENERATE  X15,FN7
      INITIAL   X11,327
      INITIAL   X15,147
      RESET     TB$TABL1-TB$TABL2
      START     1
ABLE  GENERATE  X11,FN7
BAKER GENERATE  X15,FN7
      INITIAL   X11,655
      INITIAL   X15,257
      INITIAL   X14,83
      RESET     TB$TABL1-TB$TABL2
      START     1
ABLE  GENERATE  X11,FN7
BAKER GENERATE  X15,FN7
      INITIAL   X15,514
      RESET     TB$TABL1-TB$TABL2
      START     1              END OF WEDNESDAY
ABLE  GENERATE  X11,FN7
BAKER GENERATE  X15,FN7
      INITIAL   X11,655
      INITIAL   X15,999999
      INITIAL   14,100
      RESET     TB$TABL1-TB$TABL2
      START     1
ABLE  GENERATE  X11,FN7
BAKER GENERATE  X15,FN7
      INITIAL   X15,1029
      RESET     TB$TABL1-TB$TABL2
      START     1
```

FIGURE 1-3 (contd.)

```
ABLE  GENERATE   X11,FN7
BAKER GENERATE   X15,FN7
      INITIAL    X11,327
      RESET      TB$TABL1-TB$TABL2
      START      1
ABLE  GENERATE   X11,FN7
BAKER GENERATE   X15,FN7
      INITIAL    X11,109
      INITIAL    X15,514
      RESET      TB$TABL1-TB$TABL2
      START      1
ABLE  GENERATE   X11,FN7
BAKER GENERATE   X15,FN7
      INITIAL    X15,343
      RESET      TB$TABL1-TB$TABL2
      START      1
ABLE  GENERATE   X11,FN7
BAKER GENERATE   X15,FN7
      INITIAL    X11,327
      INITIAL    X15,147
      RESET      TB$STABL1-TB$TABL2
      START      1
ABLE  GENERATE   X11,FN7
BAKER GENERATE   X15,FN7
      INITIAL    X11,655
      INITIAL    X15,257
      INITIAL    X14,83
      RESET      TB$TABL1-TB$TABL2
      START      1
ABLE  GENERATE   X11,FN7
BAKER GENERATE   X15,FN7
      INITIAL    X15,514
      RESET      TB$TABL1-TB$TABL2
      START      1               END OF THURSDAY
ABLE  GENERATE   X11,FN7
BAKER GENERATE   X15,FN7
      INITIAL    X11,783
      INITIAL    X15,999999
      INITIAL    14,100
      RESET      TB$TABL1-TB$TABL2
      START      1
ABLE  GENERATE   X11,FN7
BAKER GENERATE   X15,FN7
      INITIAL    X15,1241
      RESET      TB$TABL1-TB$TABL2
      START      1
ABLE  GENERATE   X11,FN7
BAKER GENERATE   X15,FN7
      INITIAL    X11,391
      RESET      TB$TABL1-TB$TABL2
      START      1
ABLE  GENERATE   X11,FN7
BAKER GENERATE   X15,FN7
      INITIAL    X11,130
      INITIAL    X15,621
      RESET      TB$TABL1-TB$TABL2
      START      1
ABLE  GENERATE   X11,FN7
BAKER GENERATE   X15,FN7
      INITIAL    X15,414
      RESET      TB$STABL1-TB$TABL2
      START      1
ABLE  GENERATE   X11,FN7
BAKER GENERATE   X15,FN7
      INITIAL    X11,391
```

FIGURE 1-3 (contd.)

```
      INITIAL    X15,177
      RESET      TB$TABL1-TB$TABL2
      START      1
ABLE  GENERATE   X11,FN7
BAKER GENERATE   X15,FN7
      INITIAL    X11,783
      INITIAL    X15,310
      INITIAL    X14,83
      RESET      TB$TABL1-TB$TABL2
      START      1
ABLE  GENERATE   X11,FN7
BAKER GENERATE   X15,FN7
      INITIAL    X15,621
      RESET      TB$TABL1-TR$TABL2
      START      1             END OF FRIDAY
ABLE  GENERATE   X11,FN7
BAKER GENERATE   X15,FN7
      INITIAL    X11,1180
      INITIAL    X15,999999
      INITIAL    X14,100
      RESET      TB$TABL1-TB$TABL2
      START      1
ABLE  GENERATE   X11,FN7
BAKER GENERATE   X15,FN7
      INITIAL    X15,1846
      RESET      TB$TABL1-TB$TABL2
      START      1
ABLE  GENERATE   X11,FN7
BAKER GENERATE   X15,FN7
      INITIAL    X11,590
      RESET      TB$TABL1-TB$TABL2
      START      1
ABLE  GENERATE   X11,FN7
BAKER GENERATE   X15,FN7
      INITIAL    X11,197
      INITIAL    X15,923
      RESET      TB$TABL1-TB$TABL2
      START      1
ABLE  GENERATE   X11,FN7
BAKER GENERATE   X15,FN7
      INITIAL    X15,615
      RESET      TB$TABL1-TB$TABL2
      START      1
ABLE  GENERATE   X11,FN7
BAKER GENERATE   X15,FN7
      INITIAL    X11,590
      INITIAL    X15,264
      RESET      TB$TABL1-TB$TABL2
      START      1
ABLE  GENERATE   X11,FN7
BAKER GENERATE   X15,FN7
      INITIAL    X11,1180
      INITIAL    X15,462
      INITIAL    X14,83
      RESET      TB$TABL1-TB$TABL2
      START      1
ABLE  GENERATE   X11,FN7
BAKER GENERATE   X15,FN7
      INITIAL    X15,923
      RESET      TB$TABL1-TB$TABL2
      START      1             END OF SATURDAY
      REPORT
      SPACE      6
      END
```

should be realized that every landfill operation is different. This was not considered to be a problem in this context because what is of interest here are the relative values obtained when different sampling regimens are applied, not the absolute values.

The features of the simulation model developed are as follows:

1. The landfill is operated 6 days a week (Monday through Saturday), 8 hours a day. Approximately 1000 vehicle-loads are deposited during a 1-week operation.

2. There are two classes of arriving trucks; residential loads and commercial loads. Approximately 62% of these vehicles are residential loads.

3. The residential trucks are of three types: Types 1, 2 and 3 = 20, 29, and 37 cubic yard capacity, respectively, and they are present in the system approximately in the ratio 0.07, 0.08 and 0.85, respectively. The 20 and 37 cubic yard trucks have load weight distributions which follow a triangular distribution; the load weight distribution of the 29 cubic yard trucks follows a rectangular distribution.

4. The commercial trucks are of three types: Types 4, 5 and 6 = 35, 40, 41 cubic yard capacity, respectively, and they are present in the system approximately in the ratio 0.63, 0.19 and 0.18, respectively. Their load weights all follow a rectangular distribution.

5. Approximately the last 18% of the trucks arriving at the landfill on any day contain but 83% of their normal load.

The following discussion is for those readers versed in GPSS and who may wish to use the program for their own experimentation. Other readers can proceed to Section 1.8.

The simulation program shown in Figure 1-3 prints out the following information:

1. Simulated clock time (in seconds)
2. Truck type (SAVEVALUE 29)
3. Weight (in pounds) of waste in vehicle (SAVEVALUE 30)

For the purpose of conducting sampling experiments, however, the program in Figure 1-3 is modified in three ways, the modifications consisting of removing the numbered blocks in the GPSS program and replacing them with the segment shown in Figure 1-4 and variations thereon. A sample of output for one hour is shown in Figure 1-5, consisting of four tables. Tables 1 and 2 are cumulative tables for vehicles entering the site and vehicles sampled, respectively. Tables 3 and 4 are incremental tables (i.e., statistics for that hour only) for vehicles entering and vehicles sampled, respectively. Thus, in Figure 1-5, the total number of trucks entering the landfill up to the current hour is 423, the average weight being 15125.7 pounds. The total number

```
ABLE  GENERATE   X11,FN7         GENERATE RESIDENTIAL TRUCKS
      ASSIGN     1,FN8           ASSIGN RESIDENTIAL TYPE
      TRANSFER   ,CONT
BAKER GENERATE   X15,FN7         GENERATE COMMERCIAL TRUCK
      ASSIGN     1,FN9           ASSIGN COMMERCIAL TYPE
CONT  SAVEVALUE  P1+,K1          COUNTER FOR TRUCK TYPE
      SAVEVALUE  9+,K1           COUNTER FOR SAMPLING INTERVAL
      TEST E     X9,X10,GOTO     DO WE SAMPLE THIS ONE?
      SAVEVALUE  9,K0            RESET COUNTER FOR SAMPLING INTERVAL
      SAVEVALUE  7,FN*1          FIND TRUCK WEIGHT
      SAVEVALUE  8,V$SLOW        ADJUST WEIGHT FOR END OF DAY
      PRIORITY   1               GIVE PRIORITY TO LEAVING TRUCK
      GATE SNF   WEIGH,GOTO      WEIGH STATION FULL?
      ENTER      WEIGH
      TABULATE   TABL2           TABULATE SAMPLE WEIGHTS
      TABULATE   TABL4
      ADVANCE    X12,X13         ADVANCE THROUGH WEIGHING
      LEAVE      WEIGH
GOTO  TABULATE   TABL1           TABULATE ALL TRUCKS
      TABULATE   TABL3
      TERMINATE
      GENERATE   3600        RUN FOR 1 HOUR
      TERMINATE  1



      REPORT
      SPACE      1
TAB   TITLE      ,STATISTICS FOR THIS HOUR
TAB   INCLUDE    T$TABL1-T$TABL4/1,2,3,4,5
      SPACE      3
      END
```

Figure 1-4. Sampling segment for GPSS simulation of Figure 1-3.

```
STATISTICS FOR THIS HOUR
 TABLE TABL1
 ENTRIES IN TABLE        MEAN ARGUMENT     STANDARD DEVIATION     SUM OF ARGUMENT
              423            15125.722               5344.000         6398181.000
 TABLE TABL2
 ENTRIES IN TABLE        MEAN ARGUMENT     STANDARD DEVIATION     SUM OF ARGUMENT
               57            13503.191               4976.000          769682.000
 TABLE TABL3
 ENTRIES IN TABLE        MEAN ARGUMENT     STANDARD DEVIATION     SUM OF ARGUMENT
               40            15534.472               5616.000          621379.000
 TABLE TABL4
 ENTRIES IN TABLE        MEAN ARGUMENT     STANDARD DEVIATION     SUM OF ARGUMENT
                3            12555.000               4336.000           37665.000
```

Figure 1-5. Typical simulation output, refuse truck sampling.

of trucks sampled up to the current hour is 57, the average weight being 13,503.1 pounds. (The standard deviation of the weights and sum of weights also are printed out.) For this hour only, 40 trucks entered the site (average weight 15,534.4 pounds) and 3 were sampled (average weight 12,555.0 pounds).

Figure 1-4 is shown in the Systematic sampling mode with two wheel scales. The following SAVEVALUE locations are particularly of interest:

- X10 - Sampling Interval (10 shown here),
- X9 - Random Start–a number between 1 and k for a sampling interval of k (4 shown here for a k = 10),
- X12 - Mean for weighing time, in seconds (for wheel scales, 2250 is used; for platform scales, 180), and
- X13 - Modifier for weighing time, i.e., a uniform distribution is assumed (for wheel scales, 450 is used; for platform scales, 36).

For other sampling schemes, Figure 1-4 is modified in the following ways:

(a) *Random Sampling* - Cards 7, 8 and 9 are replaced with

TRANSFER	0.900,,GOTO

This produces a 10% random sample (other sampling fractions can be produced by modifying field A of this transfer card accordingly).

(b) *A Single Scale* - Cards 13, 14 and 18 are replaced (respectively) as follows:

GATE NU	WEIGH, GOTO
SEIZE	WEIGH
RELEASE	WEIGH

(c) *Full Sampling* (i.e., sampling to the capacity of the scales) - Remove cards 7, 8 and 9.

When the sampling segment of Figure 1-4 is used, the report output of Figure 1-3 (last 3 cards) should be replaced with the report output shown at the bottom of Figure 1-4.

1.8 COMPUTER SIMULATION EXPERIMENTS

Thirteen computer runs (simulating the arrival of approximately 12,500 vehicle-loads of refuse) were made using the programs outlined in Section 1.7, the results of which are summarized in Table 1-2. "The Corrected Mean" referred to in column 3 is the mean obtained by use of Equation 1-8; the "Standard Deviation of Corrected Mean" in column 5 is the standard deviation obtained by use of Equation 1-10.

Of the thirteen runs, five of the uncorrected or "Raw" means were superior to the corrected ones. This is merely a matter of sampling error because the corrected mean is the unbiased one. To see this more clearly, we compared runs 4, 8, 9, 10, 11 and 12, which were identical except for the random number seeds used in the simulation. By changing the random number seeds in a run we simulate a different week at the site. In only two of the six runs (runs 9 and 10) were the uncorrected means superior to the corrected means. Moreover, the average absolute difference (AAD) of the raw versus true

Table 1-2. Results of Simulation Runs for One-Week Sampling Periods
(true mean = 14,760 lb/vehicle, true standard deviation = 5525 lb/vehicle, total number of trucks = 957)

Run No.	Run Description	(1) Sample Size	(2) Absolute Error of Raw Mean (lb)[a]	(3) Absolute Error of Corrected Mean (lb)	(4) Standard Deviation of Raw Mean (lb)	(5) Standard Deviation of Corrected Mean (lb)	(6) Increase in Standard Deviation of Mean (%)	(7) Loads Lost
1	10% Systematic Sample, 2 WS[c] (random start = 7)	71	120[b]	203	513	552	8	23
2	10% Systematic Sample, 2 WS (random start = 4)	74	1216	416[b]	582	627	8	22
3	10% Random Sample, 2 WS	68	175	67[b]	621	789	27	28
4	Full Sampling, 2 WS, Replication A	136	468	125[b]	416	522	25	823
5	50% Systematic Sampling, 1 PS[d]	341	221[b]	383	241	250	4	148
6	50% Random Sampling, 1 PS	279	419[b]	566	276	298	8	172
7	Full Sampling, 1 PS	429	344	138[b]	196	216	10	534
8	Full Sampling, 2 WS, Replication B	133	735	74[b]	417	545	31	855
9	Full Sampling, 2 WS, Replication C	129	240[b]	257	464	589	27	770
10	Full Sampling, 2 WS, Replication D	131	711[b]	920	436	504	16	841
11	Full Sampling, 2 WS, Replication E	136	349	175[b]	424	540	27	815
12	Full Sampling, 2 WS, Replication F	138	347	2[b]	445	546	23	870
13	Full Sampling, 1 WS	68	408	182[b]	670	897	34	892

[a]To convert lb to kg, multiply by 0.454.
[b]Estimate with the smallest bias.
[c]WS = wheel scale.
[d]PS = platform scale.

means for these six runs was 471; the AAD of the corrected versus true means for these six runs was only 209. In fact, even for all thirteen runs the comparable figures are 443 versus 270. This clearly shows the superiority of the corrected mean; it is more accurate, i.e., freer from bias.

To gain this accuracy some loss in precision is experienced (column 5 in Table 1-2), ranging between 4 to 34% increase in standard deviation of our estimate of the mean. Clearly, however, the unbiased estimates are superior in general. Even when they are not, they do not differ that much from the better estimate. The reverse cannot be said of the uncorrected estimates, as witness runs number 2, 8 and 12.

The effect of the choice of the random start in systematic sampling is shown by comparing runs number 1 and 2. The latter produced a highly biased sample (raw mean = 13,774), but the application of Equation 1-8 produced a highly satisfactory estimate (less than 3% error).

Of greater interest is the comparison between systematic and random sampling (runs 2 and 3, and runs 5 and 6). We define a "lost load" as any load that, by selection of a particular sampling plan, was scheduled to be weighed but was not because the scales were occupied at the time the vehicle arrived. (We assume that we do not hold any vehicle in a queue for weighing). In systematic sampling, the vehicles sampled are perforce "stretched out" by the sampling interval; in random sampling, the vehicles to be sampled may be "bunched up." By the laws of queuing theory, what is lost cannot be made up along the time scale. Thus, more lost loads are experienced with random than with systematic sampling.

Perhaps the most important thing to observe in Table 1-2, however, is the effect of "full" sampling, in which we sample to the fullest capacity of our scales. It is a form of the oversampling discussed in Section 1.6. Comparing run 4 with runs 2 and 3, and run 7 with runs 5 and 6, we see an appreciable *increase* in precision in full sampling over either systematic or random sampling. Since we are paying the same for the rental of the scales and the manpower, regardless of our sampling schedule, full sampling is concluded to be the optimal strategy. By contrast to the more traditional sampling plans, in which bias is minimized through the sample selection process, the methodology advocated here is one in which bias is minimized through the estimation process (i.e., Equations 1-8 and 1-11).

Runs incorporating one-platform scale (runs 5-7) and the single run (run 13) incorporating one-wheel scale need no comment because these are clearly minimum conditions for each scale type. The remainder of the runs utilized two-wheel scales because this is roughly equivalent to a single platform scale in terms of cost. At this writing, two-wheel scales would cost approximately \$700 for one week: labor, \$440 and scale rental, \$260. A one-platform scale would cost \$800: labor, \$200 and scale rental, \$600. These are rough figures, of course, that will vary according to time and location.

1.9 SEASONALITY-NUMBER OF VEHICLES APPROACH

The sampling methodology described in the previous sections are based on a sampling period of one week. Thus, hour-to-hour and day-to-day effects are accounted for in the estimating process. It is well known, however, that the quantity of municipal solid waste generated varies significantly from week to week [5]. Typically, waste generation is low during the months of January, February, November and December and peaks during June, July and August, although there are some variations depending on geographical location. Thus, it is not sufficient to sample one week out of the year to estimate generation for the complete year. Theoretically, if one knew the weeks of the year at which the curve of weekly generation crossed the horizontal line representing the average weekly generation for the year, we could schedule our one-week sampling period for one of these intersecting points and be confident that our estimate for that week, multiplied by the number of weeks in that year, would be identical to the quantity produced throughout the year. (This is termed the "critical point" approach.) This information is not available, of course. It might be available for a previous year, but there is no guarantee that the critical points will not change for the current year. Therefore, we are forced either to sample additional weeks or to use a surrogate variable. The latter will be discussed first.

From Equation 1-2 we note that the total quantity arriving at a given site can be estimated from the average load per vehicle if the total number of vehicles is known:

$$X = N\bar{X} \tag{1-2}$$

where X is the total quantity. In this case, however, N is the total number of vehicle-loads arriving per year. Our one-week sample has provided us with an estimate of $\bar{X}$; if N is counted, then X (the total quantity) can be calculated. It is true that Equation 1-2 requires that the investigator wait an entire year for N to be determined, but it is clear that the acquisition of N is relatively inexpensive may already be part of the site's routinely gathered information. Further, we may use last year's vehicle-load count with less loss in accuracy than with the critical point approach. The reason is that vehicle-load count data are commonly collected on a routine basis so sufficient data are available to estimate a trend fairly accurately. In other words, our estimate of N might be very good.

To see how errors in the estimation of N affect the estimate of X, we start with the concept of a confidence interval. Relying on the Central Limit theorem of statistics [6], a confidence interval about an estimate of a total, X, can be calculated (if the sample size is relatively large, i.e., greater than 40) as follows:

$$X \pm z\,[\mathrm{var}(X)]^{1/2} \tag{1-11}$$

where z is the standard normal deviate at some significance level, α. (For a 90% confidence interval, $z = 1.645$; for a 95% confidence interval, $z = 1.96$.) We now define "error" as one-half this confidence interval, i.e., $z\,[\mathrm{var}(X)]^{1/2}$. Since $X = N\overline{X}$, the error, E, of our estimate of X expressed as a percentage of X is given as

$$E = 100\,z\,[\mathrm{var}(X)]^{1/2}/N\overline{X}. \tag{1-12}$$

Using the propagation of error formula previously given in Equation 1-9 (however, because N is independent of $\overline{X}$, the covariance terms are zero), we find

$$\mathrm{var}(X) = N^2\mathrm{var}(X) + \overline{X}^2\mathrm{var}(N). \tag{1-13}$$

Since $X = N\overline{X}$ is nonlinear, Equation 1-13 is an approximation. Substituting Equation 1-13 into Equation 1-12, we find

$$E = 100\,z\,[N^2\mathrm{var}(\overline{X}) + \overline{X}^2\mathrm{var}(N)]^{1/2}/N\overline{X}. \tag{1-14}$$

Suppose we express the standard deviation of $\overline{X}$ as a fraction, a, of $\overline{X}$ (a is the coefficient of variation of $\overline{X}$). Then

$$[\mathrm{var}(X)]^{1/2} = a\overline{X} \tag{1-15}$$

and

$$\mathrm{var}(\overline{X}) = (a\overline{X})^2. \tag{1-16}$$

Suppose further that we express the standard deviation of N as a fraction, b, of N (b is the coefficient of variation of N). Then

$$[\mathrm{var}(N)]^{1/2} = bN \tag{1-17}$$

and

$$\mathrm{var}(N) = (bN)^2. \tag{1-18}$$

Substituting Equations 1-16 and 1-18 into Equation 1-14, we find

$$E = 100\,z\sqrt{a^2 + b^2}. \tag{1-19}$$

To see how Equation 1-19 is applied to our sampling situation, Table 1-3 has been prepared showing the percent error incurred in the estimation of X for two different values of a and three different values of b. To show a typical calculation, we compute the value in the second row of the first column of the table.

Assume that:

1. the coefficient of variation of $\overline{X}$ is 2%, i.e., $a = 0.02$;
2. the coefficient of variation of N is 1%, i.e., $b = 0.01$; and
3. we are using a 90% confidence interval, i.e., $z = 1.645$.

Table 1-3. E, Percent Error of the Total (number of vehicles approach) as a Function of Error in Estimating $\overline{X}$ and N (90% confidence level)

Coefficient of Variation of N (= b)	Coefficient of Variation of $\overline{X}$ (= a)	
	0.02	0.04
0	3.3	6.6
0.01	3.7	6.8
0.05	8.9	10.5

Then, using Equation 1-19,

$$E = 100(1.645)\sqrt{(0.02)^2 + (0.01)^2} = 3.7\%$$

The coefficients of variation of $\overline{X}$ of 2% and 4% used in Table 1-3 are roughly equivalent to what was found in the simulations of Section 1.8 with one-platform scale (full-sampling) and two-wheel scales (full-sampling), respectively. If the error in N can be kept to about 1%, Table 1-3 indicates that the error in estimating the total quantity of waste arriving at the site will be between 4 and 7%, depending on whether platform or wheel scales are used. When the error in N reaches 5%, the difference between scale types is somewhat lessened.

1.10 SEASONALITY–MULTIPLE WEEK SAMPLING APPROACH

Let us now assume that we plan to sample r weeks out of the (approximately) 52 weeks in a year and determine, for each of these r weeks, the total quantity of solid waste, X_r, arriving at the site. Equations 1-2 and 1-8 may be used for this purpose. Assume further that we obtain an average of these r totals, $\overline{Y}$. Clearly, an estimate of the total quantity for the year, Y, is given as

$$Y = 52\,\overline{Y} = 52/r \sum^{r} X_r/r \tag{1-20}$$

Also,

$$\begin{aligned} \operatorname{var}(Y) &= (52/r)^2 \sum^{r} \operatorname{var}(X_r) \\ &= (52/r)^2 \sum^{r} (s_W^2 + s_e^2) \\ &= (52)^2(s_W^2 + s_e^2)/r \end{aligned} \tag{1-21}$$

where s_W^2 is the variance due to week-to-week differences in total quantity of waste arriving at a site and s_e^2 is the error in estimating the total quantity for any given week. Let us define s_w in terms of coefficient of variation, i.e., as the fraction, c, of the weekly total, and s_e in terms of the fraction, a, of the weekly total. Note that a here is identical to that used in Equation 1-19 because the coefficients of variation are the same for totals as they are for variances [7]. Then

$$c = s_w/\overline{X} \text{ and, therefore, } s_w = c\overline{X} \tag{1-22}$$

and

$$a = s_e/\overline{X} \text{ and, therefore, } s_e = a\overline{X}. \tag{1-23}$$

Since

$$E = 100z\,[\text{var}(Y)]^{1/2}/52\,\overline{X}, \tag{1-24}$$

substituting Equations 1-21, 1-22 and 1-23 into Equation 1-24 we obtain

$$E = 100z\sqrt{(a^2 + c^2)/r}. \tag{1-25}$$

Although Table 1-2 provides us with estimates of a (e.g., rounding up to be conservative, 0.02 for a one-platform scale, full sampling, and 0.04 for two-wheel scales, full sampling), we do not as yet know c. To estimate c, we now examine some real-world data.

To this end, data were obtained for the weekly totals of solid waste delivered at the Boston municipal incinerator for the year 1962 [7] and at the Prince William County, Virginia landfill in 1975 (data obtained from the Department of Public Works, Prince William County). From these data, the coefficient of variation, c, was found to be 0.1432 at the Boston site and 0.1317 at the Prince William site.

Using the value of c obtained from the Boston data, assuming a value of a = 0.04 and supposing that 4 weeks were sampled (at random) throughout the year, the use of Equation 1-25 (at the 90% confidence level) results in an E of 12.2%. When a is assumed to be 0.02, E is 11.9%. Under these circumstances and using random sampling, a sample size of 136 from the two-wheel scales for each 1-week period is almost as effective as one of 429 from the single-platform scale. The error could be halved, of course, by sampling 16 weeks instead of 4, but this would be very costly.

Before abandoning the multiple week sampling approach in favor of the number of vehicles approach, however, it is well to review alternative plans to random sampling. To this end, another GPSS simulation program was devised that permits comparison between random and systematic sampling. The former is shown in Figure 1-6; by replacing blocks 1-13 in Figure 1-6 with the segment shown in Figure 1-7, the latter may be simulated. (The data shown in Figure 1-6 for SAVEVALUE locations X1-X52 are for Boston.) The sampling frequency for both plans was set equal to 4 ÷ 52, and each plan was simulated 1000 times (= 1000 years).

```
          SIMULATE
          INITIAL    X1,12080/X2,10280/X3,11400/X4,11600/X5,12920
          INITIAL    X6,11800/X7,10680/X8,10760/X9,11920/X10,9760
          INITIAL    X11,12520/X12,14300/X13,12400/X14,17400
          INITIAL    X15,16520/X16,17000/X17,17520/X18,17160/X19,16680
          INITIAL    X20,16900/X21,17720/X22,15960/X23,17500/X24,14720
          INITIAL    X25,17200/X26,17760/X27,15160/X28,16120/X29,15400
          INITIAL    X30,15760/X31,15480/X32,15360/X33,15160/X34,14960
          INITIAL    X35,15400/X36,15680/X37,16720/X38,15500/X39,15100
          INITIAL    X40,15000/X41,15660/X42,15240/X43,15680/X44,15360
          INITIAL    X45,15900/X46,17120/X47,15960/X48,15040/X49,16200
          INITIAL    X50,15280/X51,15380/X52,17560
     PICK FUNCTION   RN1,D52        WEIGHT DISTRIBUTION FUNCTION
    .0192,1/.0385,2/.0577,3/.0769,4/.0962,5/.1154,6/.1346,7/.1538,8
    .1731,9/.1923,10/.2115,11/.2308,12/.25,13/.2692,14/.2885,15/.3077,16
    .3269,17/.3462,18/.3654,19/.3846,20/.4038,21/.4231,22/.4423,23/.4615,24
    .4808,25/.5,26/.5192,27/.5385,28/.5577,29/.5769,30/.5962,31/.6154,32
    .6346,33/.6538,34/.6731,35/.6923,36/.7115,37/.7308,38/.75,39/.7692,40
    .7885,41/.8077,42/.8269,43/.8462,44/.8654,45/.8846,46/.9038,47/.9231,48
    .9423,49/.9615,50/.9808,51/1,52
     SELC FUNCTION   RN1,D13
    .0769,1/.1538,2/.2308,3/.3077,4/.3846,5/.4615,6/.5385,7/.6154,8
    .6923,9/.7692,10/.8462,11/.9231,12/1,13
     NORM FUNCTION   RN2,C25
    0,-5/.00003,-4/.00135,-3/.00621,-2.5/.02275,-2
    .06681,-1.5/.11507,-1.2/.15866,-1/.21186,-.8/.27425,-.6
    .34458,-.4/.42074,-.2/.5,0/.57926,.2/.65542,.4
    .72575,.6/.78814,.8/.84134,1/.88493,1.2/.93319,1.5
    .97725,2/.99379,2.5/.99865,3/.99997,4/1,5
     ADJUS FVARIABLE (FN$NORM*X60*X100)/K100+X60
     MEAN FVARIABLE  X53/4
     INCR VARIABLE   K13*X54
     CHECK TABLE     X101,1,1,54
     TABL TABLE      X55,10000,250,40
     WEEK TABLE      P1,1,1,54
          INITIAL    X100,K5
          GENERATE   ,,,4           START RANDOM SAMPLING
     LOOP ASSIGN     1,FN$SFLC      SELECT WEEK
          SAVEVALUE  60,X*1         FIND WEIGHT
          SAVEVALUE  53+,V$ADJUS    ADJUST FOR SAMPLING ERROR AND ADD
          SAVEVALUE  54+,K1         SAMPLING COUNTER
          TEST L     X54,K4,GOTO    GO TO NEXT SAMPLE?
          TERMINATE
     GOTO SAVEVALUE  54,K0          RESET COUNTER
          SAVEVALUE  55,V$MEAN      CALCULATE MEAN
          TABULATE   TABL
          SAVEVALUE  53,K0          RESET WEIGHT COUNTER
          SPLIT      4,LOOP
          TERMINATE  1
          START      1000
          END
```

Figure 1-6. Between-week refuse truck simulation of Section 1.10.

The results for varied values of a are shown in Table 1-4. Although not shown in the table, the accuracy for all methods was excellent, i.e., within 0.04–0.17% of the true mean or less than two-tenths of one percent for all runs. The result for random sampling based on the Boston data of E = 11.3% for a = 0.02 is very close to the theoretically expected value of 11.9% calculated previously. (Agreement is also excellent with the Prince William County,

```
      GENERATE    ,,,1              START SYSTEMATIC SAMPLING
LOOP  ASSIGN      1,FN$SELC         SELECT STARTING WEEK
ABLE  SAVEVALUE   60,X*1            FIND WEIGHT
      SAVEVALUE   53+,V$ADJUS       ADJUST FOR SAMPLING ERROR AND ADD
      SAVEVALUE   54+,K1            SAMPLING COUNTER
      TEST L      X54,K4,GOTO
      ASSIGN      1+,K13           NEXT SYSTEMATIC SAMPLE
      TRANSFER    ,ABLE
GOTO  SAVEVALUE   54,K0             RESET COUNTER
      SAVEVALUE   55,V$MEAN         CALCULATE MEAN OF 4 SAMPLES
      TABULATE    TABL
      SAVEVALUE   53,K0             RESET WEIGHT COUNTER
      SPLIT       1,LOOP
      TERMINATE   1
```

Figure 1-7. Systematic sampling segment for GPSS simulation of Figure 1-6.

Virginia data.) It is immediately apparent, however, that the systematic sampling plan is considerably superior to random sampling as its E is one-third to one-half as much as the latter's. This is also demonstrated by the distributions of the weight estimates for the Boston data shown in Figure 1-8. The chance for large biases is great when random sampling is used. From Section 1.9 it was found that the number of vehicles approach resulted in errors, E, of between 4 and 8%, depending on the assumptions made. Table 1-4 shows somewhat smaller errors, at least for the Boston and Prince William County data. Thus, the multiple-week-sampling approach based on systematic sampling appears to be the preferred alternative.

By conducting computer sampling experiments on the weekly data for Boston and Prince William County alone (since a is independent of c), "effective" values of c for systematic sampling can be computed. By "effective" c we mean the value of the coefficient of variation we must have in the data to achieve the same variance of the averages of size r observed in the computer sampling experiments. For quarterly sampling ($r = 4$) these values were found to be 0.0390 and 0.0513 for Boston and Prince William, respectively. For monthly sampling ($r = 12$), the values are 0.0390 (no change) and 0.0460. These values are used in the next section, which deals with sample size.

A word might be said about the characteristics of the two data sets used for the simulations of this section. The coefficient of variation, c, of the Boston data was greater than that of the Prince William County data (0.1432 versus 0.1317). For this reason, the error, E, at various values of a in Table 1-4 under random sampling was greater for the Prince William County data. Under systematic sampling, however, the situation was reversed. At $r = 4$, for example, the effective coefficient of variation was 0.0390 for

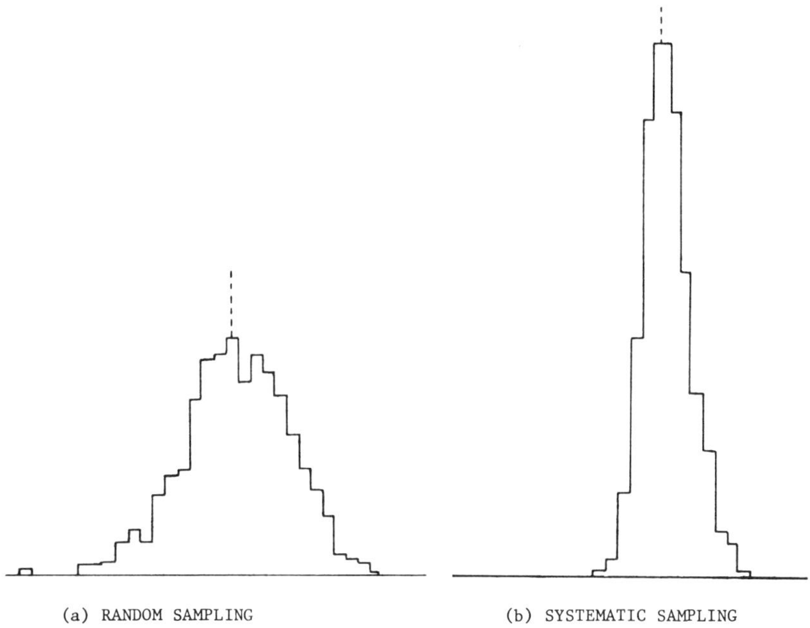

Figure 1-8. Distribution of estimates for simulations of Figures 1-6 and 1-7 at a = 0.05 (dotted lines indicate true values).

Table 1-4. E, Percent Error of the Total (multiple-week-sampling approach) as a Function of Error in Estimating $\overline{X}$ and of Type of Sampling (90% confidence level)[a]

Coefficient of Variation of $\overline{X}$ (= a)	Random Sampling (%)	Systematic Sampling (%)
Boston		
0.02	11.3	3.6
0.03	NR[b]	4.0
0.05	11.8	5.1
Prince William Co.		
0.02	10.6	5.0
0.03	10.7	5.3
0.05	11.2	6.2

[a]Each figure is an average of 1000 runs.
[b]NR = not run.

Boston and 0.0513 for Prince William County. To explain this we must introduce a method of testing data for nonrandomness [8]. This method is to find the ratio, R, of the mean square successive difference to the variance:

$$R = \frac{1/2(n-1) \sum^{n-1} (y_{i+1} - y_i)^2}{1/(n-1) \sum^{n} (y_i - \bar{y})^2} \tag{1-26}$$

It can be shown that for $n > 20$, R is approximately normally distributed with expected value, 1, and variance approximately $1/(n + 1)$. Thus, for any set of data we calculate R, then calculate the normal deviate statistic

$$z = (R - 1)\sqrt{n + 1} \tag{1-27}$$

and compare this calculated z with the desired percentage point of the normal distribution. When the data are cyclic, R will be small. For example, setting $z = 1.645$ and using Equation 1-28 we can calculate a range of R within which we would conclude that the data are truly random at the 90% confidence level. This range is $0.774 < R < 1.226$. For the Boston data, $R = 0.22$; for the Prince William County data, $R = 0.28$. Thus, both data sets are cyclic (nonrandom) in nature, as one would expect, but the Boston data are less random. Since the variance of a systematic sample decreases with decreasing randomness, the systematic sample was more effective (producing a lower value of c) in the case of the Boston data than for the Prince William County data.

In Section 1.4 it was noted that the sample size in systematic sampling will be an integer only if the population size is an integral multiple of the sampling interval. In within-week sampling, n is usually fairly large, so it makes little difference if we sample 136 trucks or 137 trucks, for example. However, r, the number of weeks sampled, is a small number, and a difference of one week from the target number can make a considerable difference. A simple solution is to consider the year to be "circular," so that the last week is followed by the first, and so on. As an example, suppose we desire to sample only 3 weeks (i.e., $r = 3$). We select a sampling interval, k, closest to the ratio $52/r$, in this case $k = 17$. Now choose a random start from 1 to 52 and apply the sampling interval, k. Suppose the random start was week 22. Then we would sample week 22, week 39 (i.e., 22 + 17), and week 4 (i.e., 39 + 17 − 52). This will result in an equal probability of selection of r weeks, each with probability $r/52$ [7].

1.11 SAMPLE SIZE DETERMINATION

A classic question in sampling is "How many samples do I need to take to estimate a quantity to within a given number of units at a given level of confidence?" This number is called the "sample size" and has nothing to do with the weight of the sample unit involved. In large sample procedures based on infinite populations where the Central Limit theorem can be expected to apply, the following formula is frequently applied to the estimation of means from simple random samples [9]:

$$n = (z\ s/d)^2 \tag{1-28}$$

where n is the sample size required, z is the standard normal deviate at a given level of significance, s is an estimate of the population standard deviation, and d is one-half the interval in which we would like our estimate of the mean to be. However, Equation 1-28 applies to infinite populations, which is not the case here because there is a finite number of vehicle-loads arriving at a given site during any single week. The easiest procedure to allow for the finite population involved is first to calculate n from Equation 1-28, then correct it by the following formula [10]:

$$n' = n/(1 + f) \tag{1-29}$$

where n' is the corrected sample size and $f = n/N_W$ and N_W is the total number of vehicle-loads, on the average, to arrive at the site during a single week. Both z and d are specified by the investigator; one must then obtain an estimate of s to utilize Equation 1-28.

Equations 1-28 and 1-29 are valid only for simple random samples, however. Because of defaulting and the use of full sampling where one undersamples at certain periods and oversamples at others, Equation 1-26 cannot be applied directly. Furthermore, it is not possible to use Equation 1-10 as a fixed condition to minimize total sample size via La Grangian multipliers or the Cauchy inequality [7], as is usually done in complex variance formula situations (the case here), because the results impose contradictory conditions on the way we are forced to, or would like to, sample. (The "optimal" solution, using these techniques, for example, is to take proportionate samples–clearly a dilemma.) However, we can use Equations 1-28 and 1-29 by incorporating the following strategem. It consists simply of computing an "equivalent" value of s, s′, then using this equivalent value in Equations 1-28 and 1-29 to determine n. By "equivalent" s we mean the value of s we would have to have to achieve with Equation 1-4 the same results of Equation 1-10, where

$$\sqrt{\text{var}(X_{ij})} = s$$

Values of s′ can be computed from Table 1-2 in lieu of other information or just plain guessing. For example, solving for $\text{var}(X_{ij})$ in Equation 1-4 we have

$$s' = \sqrt{\text{var}(X_{ij})} = \sqrt{n\ \text{var}(\bar{X})/(1 - f)} = \sqrt{n\ \text{var}(\bar{X})/(1 - n/N)} \tag{1-30}$$

Since N = 957 in Table 1-2, n is available from column 1 and var(X) is available from column 5, we can estimate s′. For full sampling, one-wheel scale (Run 13),

$$s' = 897\sqrt{68/(1 - 68/957)} = 7675$$

The values of s′ for full sampling, two-wheel scales (Run 4) and full sampling, one-platform scale (Run 7) are 6572 and 6023, respectively.

Suppose we wish to estimate $\overline{X}$ to within ±750 lb/truck (i.e., d = 750) with a confidence of 90%. This represents an error, E, of about 5%. From Equation 1-28 (assuming that full sampling, two-wheel scales are used), we obtain

$$n = (z\, s'/d)^2 = [(1.645)(6572)/750]^2 = 207.78 \text{ vehicles}$$

Assuming N = 1000, Equation 1-29 yields

$$n' = 207.78/(1 + 207.78/1000) = 172 \text{ vehicles}$$

Full sampling, two-wheel scales would not be adequate in this case since we can obtain a sample of only about 136 with such a system. However, a single-platform scale would be more than adequate.

The approach just outlined is useful, however, only if we are interested in computing the sample size necessary to estimate a mean value to within a given number of units. The mean for which the approach was applicable was the average weight of a vehicle-load for a one-week time period. For resource recovery purposes, it would be more valuable to devise a procedure applicable to the estimation of the total quantity received by a site over a one-year period. To this end, the foregoing procedure is modified as follows: To estimate the number of trucks required to be weighed during a one-week sampling period to determine the total quantity of waste delivered to a site within a one-year period using the number-of-vehicles approach, Equation 1-19 may be used together with Equation 1-15. If we rewrite Equation 1-19 with E as a fraction, q, instead of a percentage, we have

$$q = z\sqrt{(a^2 + b^2)} \tag{1-31}$$

From Equation 1-15 and expressing var($\overline{X}$) in terms of s′ we have

$$a = \sqrt{\text{var}(\overline{X})}/\overline{X} = s'/(\sqrt{n}\,\overline{X}) \tag{1-32}$$

Substituting Equation 1-32 into 1-31 and solving for n, we obtain

$$n = (z\, s'/\overline{X})^2/(q^2 - z^2 b^2) \tag{1-33}$$

with condition $q \geqslant zb$ (i.e., the error on the total cannot be less than that assumed on N). To illustrate the use of Equation 1-33 assume that the required q = 0.07 (i.e., a 7% error on the total), b = 0.01 (i.e., the coefficient of variation of N = 0.01), and z = 1.645 (i.e., 90% confidence level). For full sampling, two-wheel scales, $s' = 6572$ and, from Table 1-2, $\bar{X} =$ 14,760 lb/vehicle. Thus,

$$n = [(1.645)(6572)/14760]^2/[(0.07)^2 - (1.645)^2(0.01)^2] = 115.89$$

and assuming $N_W = 1000$,

$$n' = 115.89/(1 + 115.89/1000) = 104 \text{ vehicles}$$

To estimate the number of trucks required to be weighed during a one-week sampling period to determine the total quantity of waste delivered to a site within a one-year period using the multiple-week-sampling approach, an approach similar to that used in the derivation of Equation 1-38 may be used. We first rewrite Equation 1-25 with E as a fraction of q instead of a percentage:

$$q = z\sqrt{(a^2 + c^2)/r} \tag{1-34}$$

Substituting Equation 1-32 into 1-34 and solving for n, we obtain

$$n = (zs'/\bar{X})^2/(rq^2 - z^2c^2) \tag{1-35}$$

with condition $q \geqslant zc/\sqrt{r}$ (i.e., the error on the total cannot be less than that assumed to be due to week-to-week differences).

To illustrate the application of Equation 1-35, assume the following: $q = 0.05$, $r = 4$ weeks, $c = 0.0513$ (this is the effective c found in Section 1.10 for systematic sampling for the $r = 4$ weeks of the Prince William County data, i.e., the worst case for the two data bases), $z = 1.645$ (i.e., 90% confidence level), and $s' = 6023$ (i.e., one-platform scale, full weighing). Substituting these values into Equation 1-35, the required sample size is

$$n = [(1.645)(6023)/14760]^2/[4(0.05)^2 - (1.645)^2(0.0513)^2] = 156.53$$

Assuming $N_W = 1000$, from Equation 1-29,

$$n' = 156.53/(1 + 156.53/1000) = 135.3, \text{ or } 136 \text{ vehicle-loads}$$

Thus, two-wheel scales or a one-platform scale would be adequate for these requirements.

An interesting comparison is that of the total sample size required for a quarterly versus a monthly sampling regime. Assuming the same data as in the example just illustrated, except that now $r = 12$ and effective c (from Section 1.10) = 0.0460, from Equation 1-35,

$$n = [(1.645)(6023)/14760]^2/[12(0.05)^2 - (1.645)^2(0.0460)^2] = 18.57$$

Assuming again that $N_W = 1000$, from Equation 1-29,

$$n' = 18.57/(1 + 18.57/1000) = 18.23, \text{ or } 19 \text{ vehicle-loads}$$

Thus, for quarterly sampling, a total of 4(136) = 544 vehicle-loads is required to be weighed over the year to estimate total weight to within 5%. For monthly sampling, a total of only 12(19) = 228 are required. At first glance, monthly sampling might appear to be superior to quarterly sampling, but it is illusionary only. For one thing, we save nothing by sampling 19 vehicle-loads per week rather than 136 because the cost of scale rental and the hiring of personnel is the same in both cases. On the other hand, our scale and labor costs are tripled when we sample monthly, rather than quarterly.

Finally, a comment about the data bases used to provide the estimates of sample size illustrated here. It is perfectly true (and highly likely) that if studied, other data bases could indicate different values of the factors (s′ and effective c) than those used in the calculations of this section. If the wrong values of these factors are used, does this invalidate our estimates of parameters such as mean, total or their variances? The answer is "no" because the formulas given are valid regardless of the sample size actually used. What will change, however, is the precision of these estimates. If we take fewer samples than are needed for a given precision, then the precision we wind up with will be less than that desired. (The precisions calculated from the sample will be correct, however.) The factors given here, which are based on the sites studied by the author, are clearly useful, however, when no other information is available.

1.12 STRATIFIED SAMPLING FOR SMALL VEHICLES

Stratified sampling involves dividing the population into distinct subpopulations called strata. The strata in this case could be based on truck size, load type (residential, commercial) or any other attribute. Within each stratum a separate sample is selected and a separate estimate made. The stratum means and variances are then properly weighed to form a combined estimate for the entire population. Generally stratified sampling is used to (1) increase the precision of the estimate, (2) afford different sampling methods within the strata, and (3) provide separate estimates for different population elements. The last is of little value in the design of a resource recovery system. As for allowing different sampling methods within the strata, since large and small vehicles can be weighed on the same scales used for the one as for the other, there appears to be little incentive to try different weighing techniques in the application described in this chapter for this reason alone. However, the first may have some value where a site is used by both small private vehicles and by vehicles dedicated to waste handling. When there is a great disparity in vehicle-load weights, it makes sense to sample the lighter vehicles at a different frequency because gains in precision are achieved when the stratum elements are heterogeneous among their means [7].

It is instructive to examine the 1975 Prince William County data in greater detail. Approximately 64% of the total number of vehicles entering the site in 1975 were small, private vehicles (SPV); the remaining 36% comprised commercial vehicles (CV), "commercial" meaning those larger vehicles dedicated to waste handling. However, this 36% carried 83% of the waste (by weight) into the landfill. The number of vehicles arriving varied, of course, with the day, Saturday being the day least used by the CV, i.e., only 9.2% of the CV arrived on Saturdays. (Note: In the author's simulation of Section 1.7, based on data from the California landfill cited in the literature [4], this percentage is approximately 9.1%, obviously a very close agreement.) However, 52.3% of the SPV arrived on Saturdays. On the other weekdays, the split between SPV and CV was 48-52%, i.e., approximately equal. The site was closed on Sundays. Since the heavy arrival rate for the SPVs on Saturdays is typical, a separate sampling schedule is advisable for that day.

The sampling plan recommended is one that makes sample size proportional to the total quantity contribution of each stratum. If the standard deviation of the vehicle-loads is assumed to be proportional to the average weight of the vehicle, this is equivalent to Neyman allocation [7] and results in the following equation:

$$p_i = N_i\bar{X}_i \Big/ \sum^{2} N_i\bar{X}_i \tag{1-36}$$

where i = 1 refers to the SPV type, i = 2 refers to the CV type, p_i is the fraction of the total sample (regardless of whether the sampling is random, systematic or full) devoted to the ith type, N_i is the total number of vehicles of the ith type entering the site over a given time period, and $\bar{X}_i$ is the estimated average vehicle load weight for the ith type. An example is shown in Table 1-5.

The following standard formulas are used to estimate the means, totals and variances of the stratified sample:

$$\bar{X}_s = \sum^{2} (N_i/N)\bar{X}_i \tag{1-37}$$

$$X_s = \sum^{2} N_i\bar{X}_i \tag{1-38}$$

$$\mathrm{var}(\bar{X}_s) = \sum^{2} (N_i/N)\mathrm{var}(X_i) \tag{1-39}$$

$$\mathrm{var}(X_s) = \sum^{2} N_i^2\mathrm{var}(\bar{X}_i) \tag{1-40}$$

where $N = N_1 + N_2$ and the subscript s refers to the stratified sampling estimate. Note that if the actual sampling rates in the field differ from those

Table 1-5. Example Showing Determination of Sampling Rates in Stratified Sampling

a. *Assumptions*

SPV mean = $\bar{X}_1$ = 800 lb

CV mean = $\bar{X}_2$ = 14,760 lb

Total vehicles, Monday–Friday:

$N_1 = 840, N_2 = 910$

Total vehicles, Saturday:

$N_1 = 930, N_2 = 90$

b. *Calculations*

For Monday–Friday:

$$p_1 = \frac{(840)(800)}{(840)(800) + (910)(14760)} = 0.048$$

Therefore, for Monday–Friday, every 1/0.048 = 20.8 or 21st vehicle sampled should be an SPV.

For Saturday:

$$p_1 = \frac{(930)(800)}{(930)(800) + (90)(14760)} = 0.359$$

Therefore, on Saturday, every 1/0.359 = 2.8 or 3rd vehicle sampled should be an SPV.

---

calculated from Equation 1-36, the above estimates are still valid because Equations 1-37 and 1-40 are based on the actual sampling rates. The only thing lost is some sampling efficiency, i.e., the variances calculated using Equations 1-39 and 1-40 would be higher than if the sampling fractions computed from Equation 1-36 had been achieved.

Of course, Equations 1-8 and 1-10 are used to estimate the $\bar{X}_i$ and var($\bar{X}_i$), respectively, in Equations 1-37 to 1-40, bearing in mind that when so used, the n_i, m_i and m of Equations 1-8 and 1-10 refer to one specific vehicle type only (SPV or CV), as if the other type did not exist. If the sampling plan devised in Table 1-5 had been implemented and Equation 1-8 estimated $\bar{X}_1$ as 810 lb/load and $\bar{X}_2$ as 14,300 lb/load; our stratified mean would be (using Equation 1-37)

$$\bar{X}_S = \frac{(1770)(810) + (1000)(14300)}{1770 + 1000} = 5680 \text{ lb/vehicle}$$

Note again that if our actual sampling rates had differed from those calculated in Table 1-5 (i.e., every 21st vehicle Monday-Friday and every 3rd vehicle on Saturday being an SPV), this estimate of 5680 lb/vehicle would still be valid.

1.13 SUMMARY

The sampling plans described in this chapter are tailored specifically to the sampling of raw municipal solid waste delivered to a disposal or transfer site. Whether scales are available or must be rented, the sampling plans are designed to utilize these scales to the fullest extent to maximize efficiency. This means biasing the sample, but the bias is removed in the process of estimation by the formulas presented. The concept of full sampling has a further advantage—that of simplicity. There is no sampling plan that conceivably could be easier to implement in practice.

The general problem of seasonality has been overlooked in the solid waste literature in the past, but has been treated here in much detail, along with the special, but very real, problems that arise when many small, private vehicles frequent the site. Sample size has also been discussed based on the data bases studied by the author, but whatever the sample size actually used, all estimates calculated by the equations provided are valid and unbiased. A great deal of care has been given to the design of sampling plans that incorporate the seasonality effect and are also economical and practical.

Finally, after accounting for within-day, between-day and between-week variability in solid waste quantity by the use of the sampling methodology outlined in this chapter, all that remains to fully design a resource recovery facility in terms of estimation of quantity is the problem of future changes in the quantity of waste produced. This is the subject of Chapter V, and the forecasting techniques described there are designed to help complete the total picture of estimating solid waste quantity for resource recovery purposes.

APPENDIX TO CHAPTER 1

A 1.1 PROOF THAT $\overline{X}$ IS UNBIASED

To show that Equation 1-8 is an unbiased estimate of the true mean, u, let u_i be the true mean in the ith time period of sampling, T_i be the true total for the ith period, and T be the total for the entire population consisting of H time periods. Then, from Equation 1-8 and taking the expectation, E, of both sides,

$$E(X) = \sum_{i}^{H} \frac{m_i}{m} E \sum_{j}^{n_i} X_{ij}/n_i$$

$$= \sum_{i} \frac{m_i}{m} u_i = \sum_{i} T_i/m = T/m = u$$

A 1.2 DERIVATION OF VAR($\overline{X}$)

Let a_{ij} be a random variate that takes the value 1 if the jth unit is in the sample and the value 0 otherwise. The sample mean of Equation 1-8 may be written as follows:

$$\overline{X} = \sum_{i}^{H} \sum_{j}^{m_i} \frac{m_i}{mn_i} a_{ij} X_{ij} \tag{A-1}$$

where H is the total number of time periods over the duration of sampling. This sum extends over all the N units in the population because

$$\sum_{i}^{H} m_i = N \tag{A-2}$$

Clearly, for a given i,

$$\Pr(a_{ij} = 1) = \frac{n_i}{m_i}, \text{ and } \Pr(a_i = 0) = 1 - \frac{n_i}{m_i}$$

Thus, a_{ij} is distributed as a binomial variate in a single trial, with $P = n_i/m_i$. Hence

$$E(a_{ij}) = P = n_i/m_i, \text{ and } \operatorname{var}(a_{ij}) = PQ = \frac{n_i}{m_i}\left[1 - \frac{n_i}{m_i}\right] \tag{A-3}$$

To find var($\overline{X}$) we need also the covariance of a_{ij} and a_{ik}. The product $a_{ij}a_{ik}$ is 1, if both subscripted units are in the sample and is zero otherwise. The probability that two specific units are both in the sample for any given i is clearly $n_i(n_i - 1)/m_i(m_i - 1)$. Hence,

$$\operatorname{cov}(a_{ij}a_{ik}) = E(a_{ij}a_{ik}) - E(a_{ij})E(a_{ik}) = \frac{n_i(n_i - 1)}{m_i(m_i - 1)} - \left[\frac{n_i}{m_i}\right]^2 = -\frac{n_i}{m_i(m_i - 1)}\left[1 - \frac{n_i}{m_i}\right] \tag{A-4}$$

Applying the propagation of error concept [2] to find var($\overline{X}$) and setting, for convenience, $Y_{ij} = m_i X_{ij}/n_i$, we have from Equation A-1

$$\operatorname{var}(X) = \frac{1}{m^2}\left[\sum_{i}^{H} \sum_{j}^{m_i} Y_{ij}^2 \operatorname{var}(a_{ij}) + \sum_{i} \sum_{j \neq k} Y_{ij} Y_{ih} \operatorname{cov}(a_{ij}a_{ih})\right]$$

$$= \frac{1}{m^2}\left[\sum_{i}^{H} \sum_{j}^{m_i} Y_{ij}^2 \frac{n_i}{m_i}(1 - f_i) - \sum_{i} \sum_{j \neq k} Y_{ij} Y_{ih} \frac{n_i}{m_i(m_i - 1)}(1 - f_i)\right]$$

$$= \frac{1}{m^2}\left[\sum_{i}^{H} \frac{n_i}{m_i}(1 - f_i)\left(\sum_{j} Y_{ij}^2 - \frac{1}{m_i - 1}\sum_{j \neq k} Y_{ij} Y_{ih}\right)\right] \tag{A-5}$$

where $f_i = n_i/m_i$. Completing the square on the cross-product term gives

$$\mathrm{var}(X) = \frac{1}{m^2}\left[\sum_i^H \frac{n_i}{m_i}(1 - f_i)\left(\frac{m_i}{m_i - 1}\sum Y_{ih}^2 - \frac{1}{m_i - 1}Y_i^2\right)\right] = \frac{1}{m^2}\sum_i^H n_i(1 - f_i)\mathrm{var}(Y_i) \tag{A-6}$$

But $\mathrm{var}(Y_i) = m_i^2\mathrm{var}(X_{ij})/n_i^2$. Thus,

$$\mathrm{var}(\overline{X}) = \frac{1}{m^2}\mathrm{var}(X_{ij})\sum_i^H m_i^2(1 - f_i)/n_i \tag{A-7}$$

A 1.3 PROOF OF EFFICIENCY OF OVERSAMPLING

To show that oversampling is more efficient (i.e., has smaller variance) than unbiased systematic or simple random sampling, it is sufficient to show that the coefficient of the population variance in Equation 1-4 is greater than that in Equation 1-10, i.e.,

$$(1 - f)/n \geqslant (1/m^2)\sum m_i^2(1 - f_i)/n_i \tag{A-8}$$

Since $m = \sum m_i$, $f_i = n_i/m_i$ and $f = n/m$, inequality A-8 can be rewritten (after first multiplying both sides by m^2) as follows:

$$(m^2 - m^2n/m)/n \geqslant \sum (m_i^2 - m_i^2 n_i/m_i)/n_i$$

So that

$$m^2/n - m \geqslant \sum m_i^2/n_i - \sum m_i$$

and

$$m^2/n \geqslant \sum m_i^2/n_i \tag{A-9}$$

Thus it is necessary only to show that inequality A-9 holds.

We start by noting that in systematic sampling (assuming that the n_i' are integral multiples of k), $n_i' = m_i/k$ and $k = m/n$, where k is the sampling interval. (Since our symbol for the number of trucks sampled in the ith hour is n_i for biased sampling, we have used n_i' for its unbiased counterpart.) If the n_i' are not integral multiples of k or if simple random sampling is used, then the n_i' will not necessarily be equal to m_i/k. However, the expectation of n_i', $E(n_i')$, is equal to m_i/k in such cases. When sampling to the fullest capacity of our scales, n_i cannot be less than n_i', so $E(n_i') \leqslant n_i$ for all i. Consequently,

$$m^2/n = mk = \sum m_ik = \sum m_i^2/(m_i/k) = \sum m_i^2/E(n_i') \geqslant \sum m_i^2/n_i$$

and inequality A-9 is proved.

REFERENCES

1. Kish, L. "Introduction," in *Survey Sampling* (New York: John Wiley & Sons, Inc., 1965), pp. 3-34.
2. Deming, W. E. "Some Variances in Random Sampling," in *Some Theory of Sampling* (New York: John Wiley & Sons, Inc., 1950), pp. 127-134.
3. Schriber, T. J. *Simulation Using GPSS* (New York: John Wiley & Sons, Inc., 1974), pp. 1-533.
4. Woodyard, J. "Municipal Solid Waste Survey Protocol," Contract No. 68-03-2486, U.S. Environmental Protection Agency, SCS Engineers, Inc., Long Beach, CA (1979).
5. "Waste Flow Characteristics and Resource Recovery," Contract No. E(11-1)-4011, Division of Buildings and Community Systems, Office of Conservation, Energy Research and Development Administration, Resource Planning Association, Inc., Washington, DC (October 8, 1976), pp. A 1-28.
6. Fraser, D. A. S. "Sampling from Probability Distribution," in *Statistics, An Introduction* (New York: John Wiley & Sons, Inc., 1958), pp. 121-123.
7. Harrington, J. J., and L. J. Partridge, Jr. "Analysis of Variability in Loadings on a Municipal Refuse Incinerator," *J. Air Poll. Control Assoc.* 23(4):282-286 (1973).
8. Smith, H. "A Simple Method of Testing for Nonrandomness of Data," *Ind. Qual. Control* 22(3):691-692 (1961).
9. Mace, A. E. "Estimation Problems," in *Sample Size Determination* (New York: Van Nostrand Reinhold Co., 1964), pp. 34-72.
10. Kish, L. *Survey Sampling* (New York: 1964), John Wiley & Sons, Inc., 1965.

CHAPTER 2

ESTIMATION OF WASTE COMPOSITION

2.1 INTRODUCTION

The estimation of waste composition in some respects is a more difficult task than the estimation of waste quantity for at least three reasons:

1. Complexity: Unlike waste quantity, waste composition involves the measurement of more than one characteristic. Depending on the nature of the information required, one may be measuring from four to eleven components of the waste stream.
2. Cost: Weighing a collection vehicle is a relatively low-cost procedure. Selecting a sample of waste and separating it into a number of components prior to weighing is much more expensive.
3. Lack of data: Because of the time and expense required to sample for waste composition, there are fewer data available regarding this aspect of waste characterization than for the estimation of waste quantity.

Finding procedures for the selection of samples and estimating sample size are two problems common to both composition and quantity estimation. (By "sample size" we mean the *number* of sample units to be separated, not the *weight* of each sample unit.) One problem that is unique in composition estimation, however, is the definition of the sample unit. The sample unit in quantity estimation clearly is the vehicle. One weighs the entire vehicle because it makes no sense to select and just weigh a portion of a vehicle-load. In composition sampling, however, we usually cannot separate an entire vehicle-load because the net weight of the usual commercial vehicle load typically weighs between 4500 and 9000 kg (10,000 and 20,000 lb). This is obviously too large a sample unit to separate. (It must be remembered that separation is done manually. As a very rough approximation, typically one man can separate 30–140 kg (65–300 lb) of municipal solid waste in one hour, depending on the number and type of components desired.)

Clearly, very small sample weights make no sense physically; a large piece of metal or wood, for example, could not physically ever be included in a sample of, say, 2.3 kg (5 lb). Furthermore, small sample weights tend to be more homogeneous than the population being sampled, i.e., the smaller the sample weight, the greater the likelihood that it consists entirely of wood or of paper or of glass, etc. Following a well-known principle of cluster sampling [1], this homogeneity tends to increase the variance of the sample. Indeed, Klee and Carruth [2] found that the smaller the sample weight, the greater the variance of the municipal solid waste sample. However, the relationship was not linear. Under 90 kg (200 lb) the sample variance increased rapidly; over 140 kg (300 lb) it increased much more slowly. Accordingly, they recommended that a sample weight of 90–140 kg (200–300 lb) generally be used for raw refuse sampling, and this recommendation has been widely adopted by other investigators for the field sampling of raw refuse. (The determination of sample weight for process streams, i.e., intermediate streams in a resource recovery plant, is discussed in Section 2.5.)

Like vehicle-load weight, the composition of municipal solid waste can be expected to vary with time of day and day of week; therefore, it is important to sample throughout the day and the week. Because the determination of composition in a waste sample is a labor-intensive, potentially dangerous and lengthy process, it is not practicable to obtain as many samples for composition as it is for quantity determination. Thus, the taking of composition samples is not likely to result in vehicle queues. A sample can be taken from a selected vehicle and the vehicle allowed to move on. For this reason, an unbiased sample is recommended, e.g., either the random or systematic sample described in Chapter 1. In other words, once sample size has been determined, the actual sample selection is based on either of the two techniques discussed in Section 1.4.

2.2 SAMPLE SIZE: BASIC CONSIDERATIONS FOR WITHIN-WEEK SAMPLING

The determination of sample size for the within-week estimation of composition is, unfortunately, a much more complex matter than for quantity estimation. This section focuses only on the within-week case—estimation of composition for a one-week period. Of course, actual composition sampling for a resource recovery system must take seasonality into consideration, and this is dealt with in Section 2.4. The underlying theoretical distribution of solid waste composition in a sample is the multinomial distribution. Unfortunately, the number of samples required to estimate the percentage of each component in the population is rather high when the underlying distribution is multinomial. The exact procedure is outlined by Tortora [3], the relationship being as follows:

$$n = BP(1 - P)/d^2 \qquad (2\text{-}1)$$

where n is the sample size, P is the true fraction of the component for which we are estimating sample size, d is the precision required, and B is the upper $(\alpha/k) \times 100$th percentile of the chi-square distribution with 1 degree of freedom, where α is the desired significance level and k is the number of components of interest in the population. (Because the number of 90- to 140-kg (200- to 300-lb) samples that practically can be taken is a very small fraction of the total weight of the waste arriving at a site, no finite population correction factor is needed for any of the formulas in this chapter.) As an example, suppose we wish to estimate the percentage of metals in raw solid waste to within two percentage units with a confidence of 90%. Assume that metals are expected to be found at a concentration of approximately 9%. We assume further that there are seven components of interest. Thus:

$$P = 0.09$$

$$d = 0.02$$

$$\alpha = 0.10$$

$$k = 7$$

$$\alpha/k = 0.0143$$

$$1 - \alpha/k = 0.9857$$

Chi-square at 0.9857 probability (chi-square tables must be interpolated to find this number) and 1 degree of freedom = 6.58. Using Equation 2-1,

$$n = BP(1 - P)/d^2 = 6.58(0.09)(0.91)/(0.02)^2 = 1347$$

Thus, 1347 samples of 90–140 kg (200–300 lb) each are required, clearly not a practical condition. An alternative, outlined by Cochran [4], is to consider each component versus the remaining components as a binomial distribution and thus obtain a set of binomial sample size estimates for the individual cell proportions. The difficulty here is that we do not know the confidence level for the set of components as a whole. We know it only for the individual components. The author does not see this as a problem because the selection of a confidence level is arbitrary (in classical Neyman–Pearson statistics, at least). The appropriate formula is as follows:

$$n = t^2P(1 - P)/d^2 \tag{2-2}$$

where t is the t-value at the required confidence level. Since t is dependent on n, one must use a method of trial and error in which the t-value in Equation 2-2 is replaced in subsequent trials. (This will be illustrated later in this section.) As a first approximation to t, we can use the standard normal variate, z, at the required confidence level. At the 90% level, z = 1.645. Redoing the aforementioned problem, the use of Equation 2-2 results in

$$n = (1.645)^2(0.09)(0.91)/(0.02)^2 = 554$$

Since, at n = 554, the t-value is almost identical with the standard normal deviate, z, trial and error is not required here and the calculation is complete. The sample size of 554 as found here is also impractical, however.

Klee and Carruth [2] approached this problem by using the normal distribution approximation of Equation 1-29, which is slightly modified here as

$$n = (ts'/d')^2 \tag{2-3}$$

where the primes represent transformed values. To permit the use of Equation 1-29 of Chapter 1, a normalizing transformation must be used. Klee and Carruth used the arcsin (or inverse sine) transformation. This is appropriate to the bionomial distribution so the confidence levels selected are applicable to each component estimated, not to the set of estimates as a whole. The procedure is as follows: first we calculate the precision required on a transformed basis using

$$d' = 2\,(\arcsin \sqrt{P \pm d}) - 2\,(\arcsin \sqrt{P}) \tag{2-4}$$

The choice of sign for $P \pm d$ is determined as follows: positive if $P < 0.50$, and negative if $P \geqslant 0.50$.

For example, in our problem,

$$d' = 2\,(\arcsin \sqrt{0.09 + 0.02}) - 2\,(\arcsin \sqrt{0.09})$$
$$= 2\,(0.3381) - 2\,(0.3047) = 0.0668$$

where the arcsin values were obtained from Weast [5]. To use Equation 2-3, an estimate of s′ must be available. Klee and Carruth [2], based on extensive composition data from seven sites, suggested a value of 0.1632 for 90- to 140-kg (200- to 300-lb) samples. As a first approximation we use z in place of t, and using Equation 2-3 we find

$$n = [(1.645)(0.1632)/0.0667]^2 = 16.2 \text{ or } 17 \text{ samples}$$

However, at n = 17, the t-value is greater than 1.645 and it is clear that a slightly higher sample size is required. Assuming that n = 18, t = 1.74 [5] and Equation 2-3 yields

$$n = [(1.74)(0.1632)/0.0667]^2 = 18.13 \text{ or } 19 \text{ samples}$$

At n = 19, t = 1.734, and

$$n = [(1.734)(0.1632)/0.0667]^2 = 18.001 \text{ or } 19 \text{ samples}$$

Therefore, n = 19 is correct. Clearly a sample size of 19 is a much more practical condition.

Further reductions in sample size can be made if we can assume that the components are normally distributed *without* prior transformation. If the assumption is valid, then the sample size is given by [6]:

$$n = (ts/d)^2 \qquad (2\text{-}5)$$

Britton [7] shows that, for 90-kg (200-lb) samples, food waste, paper, metal and glass/ceramics are distributed normally. For 140-kg (300-lb) samples, all categories studied by him but textiles and wood were found to be distributed normally. However, the distribution of the basic sample itself is, in one sense, somewhat irrelevant for the purpose of determining sample size. In a one-week sampling period we are interested in determining *average* compositions and, as is well known, the Central Limit Theorem referred to in Chapter 1 suggests that these means might be normally distributed.

First, however, let us establish a motivation for the use of Equation 2-5 by applying it to our metals problem. An estimate of s = 0.0358 is provided in Table 2-1. (These estimates are based on various composition sampling studies throughout the country and include within-day and between-day sampling variation. The Britton estimates are *not recommended* because they represent only the within-truck variation of one truckload.) We start as before, assuming t = 1.645 (90% confidence level and n = infinity):

$$n = [(1.645)(0.0358)/0.02]^2 = 8.7 \text{ or } 9$$

The t-value at n = 9 is much greater than 1.645. At n = 11, t = 1.812 and

$$n = [(1.812)(0.0358)/0.02]^2 = 10.52 \text{ or } 11$$

Table 2-1. Suggested Standard Deviations, s, for Within-Week Variation of Raw Solid Waste Components for 90- to 140-kg (200- to 300-lb) Samples

Components	Standard Deviation, s
Food Waste	0.0506
Garden Waste	0.1629
Total Paper	0.1021
Corrugated Paper	0.0744
Newsprint	0.0687
Plastic/Rubber/Leather	0.0252
Textiles	0.0687
Wood	0.1376
Total Organics	0.1121
Total Metals	0.0358
Ferrous Metals	0.0388
Aluminum	0.0069
Glass/Ceramics	0.0502
Ash/Rocks/Fines	0.0572

Since our estimated n is now in accordance with its t-value in Equation 2-5, we are finished, and the estimated sample size, 140-kg (300-lb) samples, is 11. This is even a greater improvement over the required sample size of 19 determined by the use of Equation 2-3. Thus, if it is valid, the use of Equation 2-5 has decided advantages. The validity of the equation is shown in the next section.

2.3 COMPUTER SAMPLING EXPERIMENTS TO TEST NORMALITY ASSUMPTIONS

Seven municipal solid waste components were selected for simulation:

- Newsprint
- Total Paper
- Plastics/Rubber/Leather
- Textiles
- Wood
- Ferrous Metals
- Garden Waste.

Basic (population) distributions of the components are needed for computer sampling studies, and the distributions used were based on the original data collected by Britton [7]. The population distributions are shown in the n = 1 histograms shown in Figures 2-1(a)–2-1(g), and all histograms represent distributions of 1000 random values each. None of the seven population distributions is normally distributed; indeed, the distributions for textiles, wood and garden waste are highly skewed.

Random sampling from these distributions was then performed using a digital computer (and GPSS), which generated distributions of averages of 4, 10 and 20 samples. (For wood and textiles, averages of 35 and 50 were also generated.) For averages of 4, Figure 2-1 clearly shows that newsprint, total paper, plastics/rubber/leather and ferrous metals are normally distributed; such is the power of the Central Limit theorem in statistics. However, garden waste approaches normality only at n = 10, and both wood and textiles still are not quite normally distributed at n = 35 and 50, respectively, although for practical purposes the normal distribution would be quite adequate for resource recovery estimation purposes.

That this is so is seen from Table 2-2, which shows the sample size needed (90% confidence) to estimate these components to within 20% of the mean. (Equation 2-5, the standard deviations shown in Table 2-1 and the means observed by Britton were used to calculate these sample sizes.) At the sample sizes required for these components, all of the averages would be distributed normally. We have shown, therefore, that normal theory (Equation 2-5) is valid for the type of composition sampling likely to be encountered in actual practice.

Table 2-2. Sample Size Required to Determine Selected Solid Waste Components to Within 20% of the Mean (90% confidence and one-week generation period)

Component	Required Sample Size
Newsprint	45
Total Paper	8
Plastics/Rubber/Leather	27[a]
Textiles	275
Wood	724
Ferrous Metals	31
Garden Waste	139

[a]Example of Calculation

For Plastics/Rubber/Leather, s = 0.0252 from Table 2-1. We have assumed $d/\bar{X} = 0.20$, and from Britton [7], $\bar{X} = 0.042$. Using as a first approximation z = 1.645, Equation 2-5 shows

$$n = \left[\frac{(1.645)\,(0.0252)}{(0.20)\,(0.042)}\right]^2 = 24.35 \text{ or } 25$$

We guess that n = 27 might be sufficient. At n = 27, t = 1.706 and

$$n = \left[\frac{(1.706)(0.0252)}{(0.20)(0.042)}\right]^2 = 26.2 \text{ or } 27$$

Since the assumed n now equals the calculated n, n = 27 is the correct sample size.

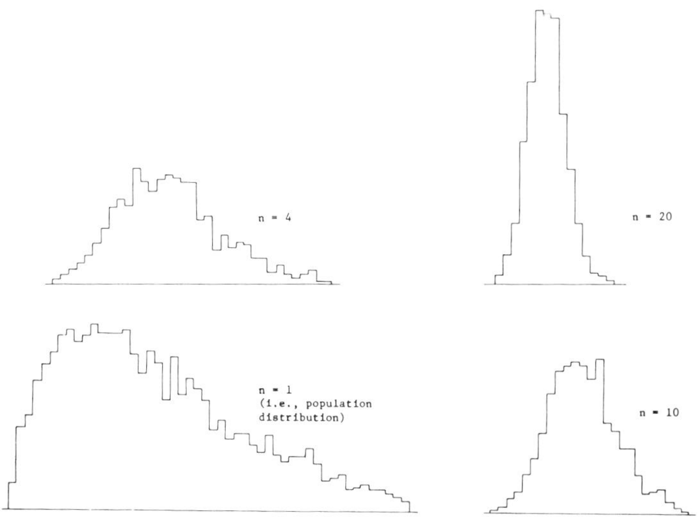

Figure 2-1(a). Distribution of averages–newsprint.

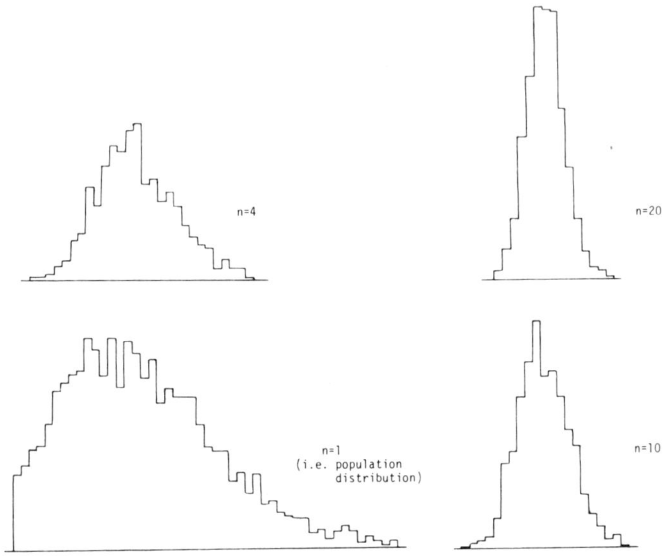

Figure 2-1(b). Distribution of averages–total paper.

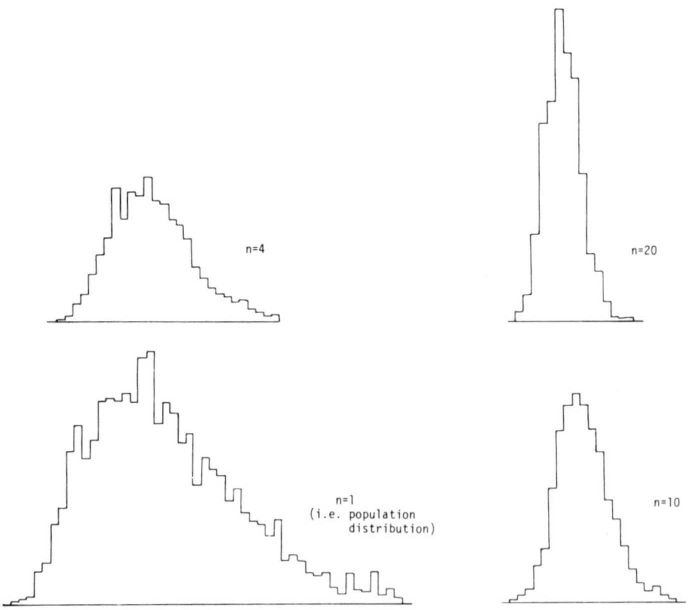

Total 2-1(c). Distribution of averages–plastics/rubber/leather.

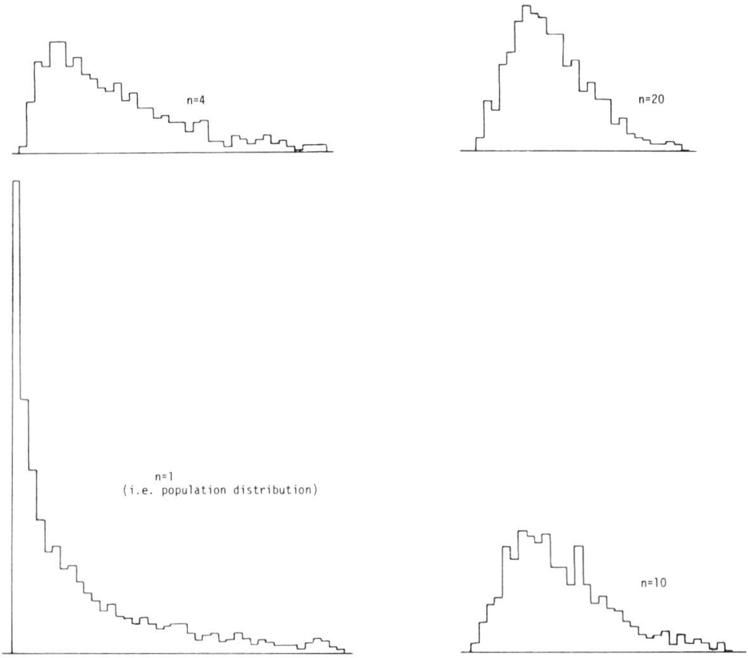

Total 2-1(d). Distribution of averages—textiles.

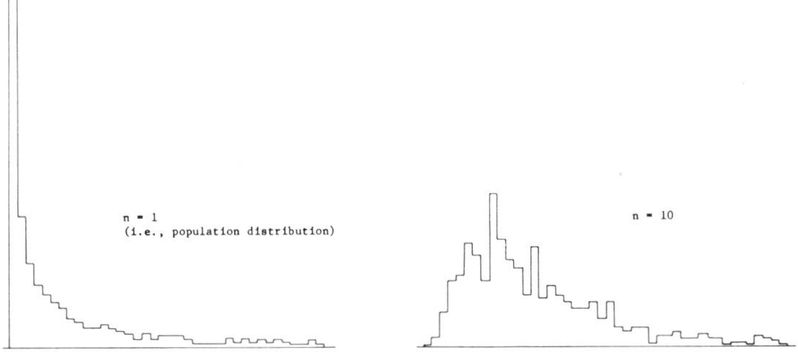

Figure 2-1(e). Distribution of averages—wood.

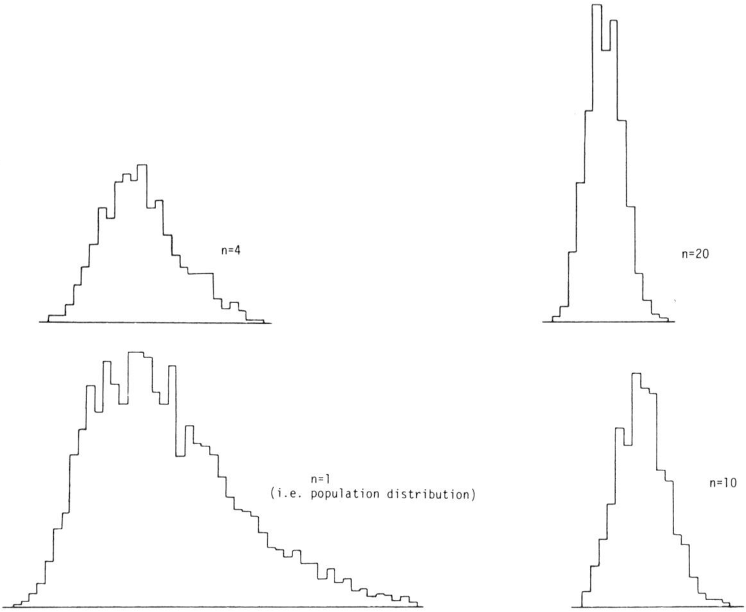

Figure 2-1(f). Distribution of averages–ferrous metals.

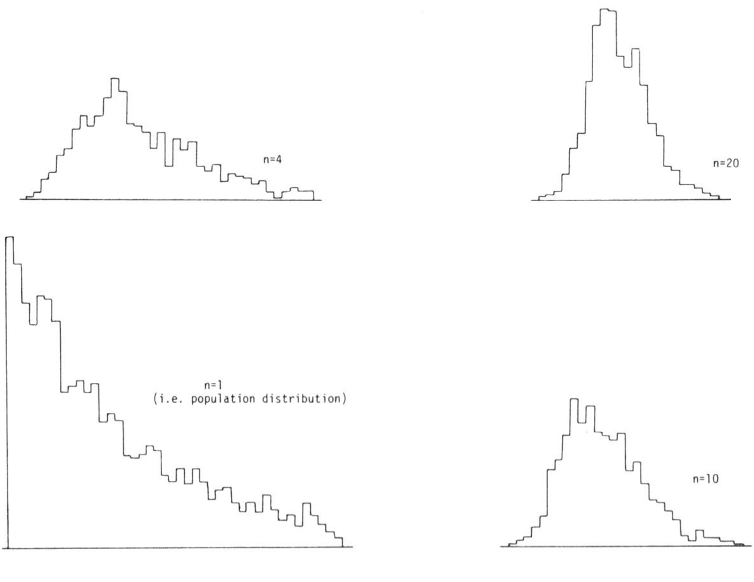

Figure 2-1(g). Distribution of averages–garden waste.

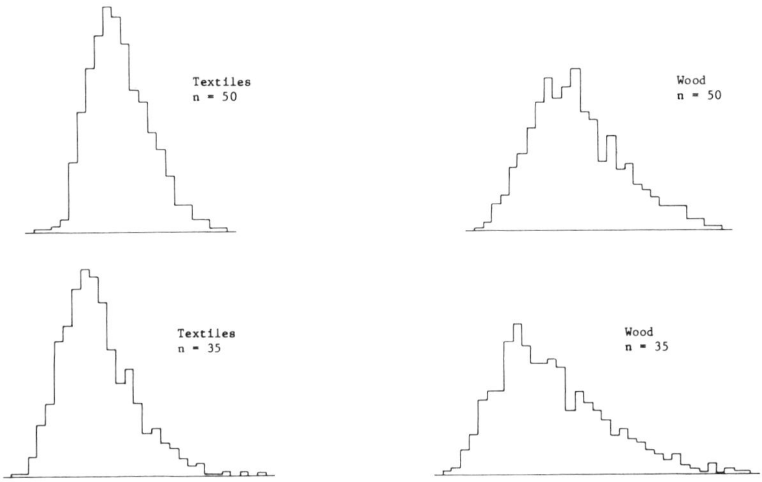

Figure 2-1(h). Distribution of averages–wood and textiles.

2.4 SAMPLE SIZE WITH SEASONALITY

In Section 2.2 it was assumed that sampling for composition would be based on a sampling period of one week. As with quantity, however, the composition of municipal solid waste also varies significantly from week to week. We now assume that we plan to sample r weeks out of the (approximately) 52 weeks in a year to make an estimate of average yearly composition. For each of these r weeks, we determine the average composition, $\bar{c}_i$, by

$$\bar{c}_i = \sum_{i=1}^{n} c_i/n \tag{2-6}$$

where c_i is the fraction of a given component in the ith sample of 90–140 kg (200–300 lb), and n is the total number of samples obtained in each one-week period. (Note that n is assumed to be the same for all r weeks. Since there is more than one component of interest, the c symbol should be double-subscripted, i.e., c_{ij}, where j represents the jth component, but as this merely complicates the equations, it is omitted here for simplicity.) An estimate of the average yearly composition, C, is

$$C = \sum_{t=1}^{r} \bar{c}_t/r \tag{2-7}$$

$$\text{var}(C) = (1/r)^2 \sum^{r} \text{var}(\bar{c}_t)$$

$$= (s_W^2 + s^2/n)r \tag{2-8}$$

where s_W^2 is the variance due to week-to-week differences in composition and s is the within-week error (i.e., the values given in Table 2-1). Since a one-half confidence interval, d, is given as

$$d = t\sqrt{\text{var}(C)} \tag{2-9}$$

by substituting Equation 2-8 into Equation 2-9 and solving for n, we obtain

$$n = (ts)^2/(d^2r - s_W^2t^2) \tag{2-10}$$

with condition $d \geqslant s_W t/\sqrt{r}$ (i.e., the one-half confidence interval must be at least equal to the result if the within-week error were zero).

Thus, to determine the number of samples required to estimate a component of municipal solid waste to within a desired interval that incorporates seasonality (week-to-week effects) as well as hour-to-hour and day-to-day effects, Equation 2-10 is used. It is necessary, however, to have an estimate of s_W to use this equation. To this end, data were obtained for the monthly composition of various waste components observed in the city of Montreal from June 1970 to May 1971 [8]. For each component, a smooth curve of its fraction of the total was plotted, and weekly values were taken from the curve. Analogously to the treatment in Section 1.10, we compare simple random sampling with systematic sampling using a computer simulation approach. However, one additional sampling plan is added, i.e., systematic sampling with changed start. This method is similar to systematic sampling (which was discussed in Section 1.4) in that the population is stratified into r intervals of k units each, but differs in that we choose a different random number between 1 and k in each interval. For example, if we assume a 52-week year and r = 4, then k = N/r = 52/4 = 13. Let us further assume that the four random numbers between 1 and 13 selected from a table of random numbers are 5, 9, 8 and 11. The weeks to be sampled then are 5, 9 + 13, 8 + 2(13) and 11 + 3(13), or weeks 5, 22, 34 and 50.

The results of these simulations are shown in Table 2-3. First, let us direct our attention to the first three columns of the table (i.e., quarterly sampling or r = 4). As was found in Section 1.10, systematic sampling was generally superior to simple random sampling. However, in three instances (total paper, newsprint and total organics), systematic sampling resulted in a slightly higher standard deviation, s_W, than did simple random sampling (about 16% greater for each of these three components). However, systematic sampling with changed start was superior to simple random sampling in all cases. Further, between the two types of systematic sampling, they were roughly equal in three components (plastics/rubber/leather, textiles and aluminum); systematic with changed start was better on six components (total paper, total metals, ferrous metals, ashes/fines, total organics and newsprint); and ordinary systematic was superior on five components (food waste, garden wastes, wood, glass/ceramics and corrugated paper). Since total organics, ferrous metals, aluminum and glass/ceramics are the most important components to be estimated for resource recovery purposes, systematic sampling with changed start is the preferred method for general purposes.

Table 2-3 also presents the results of the simulation runs of systematic sampling with changed start for monthly and biweekly sampling ($r = 12$ and $r = 26$, respectively) as well as quarterly ($r = 4$) sampling. Since the week is the basic unit of observation for s_W and since there is a small, finite number of possible observations (i.e., 52), s_W changes with r. This is analogous to the change of c with r, discussed in Section 1.10 of Chapter 1. Intermediate values of r can be obtained from the table by interpolation.

An example of the approach to sample size determination first developed in Section 2.2 and this section now follows. We assume that we desire to estimate the percentage of total organics, ferrous metals, aluminum and glass/ceramics to within 6, 2, 0.2 and 2 percentage points, respectively. The confidence level desired is 90%, and it is planned to take four systematic samples with changed starts throughout the year. How many samples of 90–140 kg (200–300 lb) do we have to take in each of the 4 weeks? We apply Equation 2-10 to each of the components using the values of s and s_W given in Tables 2-1 and 2-3, respectively, and using the value 1.645 as a first approximation to t at 90% confidence:

$$\text{TO: } n_1 = [(1.645)(0.1121)]^2/[(0.06)^2(4) - (0.0312)^2(1.645)^2] = 2.9$$

$$\text{Fe: } n_2 = [(1.645)(0.0388)]^2/[(0.02)^2(4) - (0.0171)^2(1.645)^2] = 5.1$$

$$\text{Al: } n_3 = [(1.645)(0.0069)]^2/[(0.002)^2(4) - (0.002)^2(1.645)^2] = 24.9$$

$$\text{G/C: } n_4 = [(1.645)(0.0502)]^2/[(0.02)^2(4) - (0.0144)^2(1.645)^2] = 6.6$$

Since the largest n is that for aluminum, this component is the controlling component in the calculation of n; consequently, it is not necessary to consider the other components further. At $n = 31$, $t = 1.697$. Therefore,

$$n_3 = [(1.697)(0.0069)]^2/[(0.002)^2(4) - (0.002)^2(1.697)^2] = 30.6 \text{ or } 31$$

Thus we would sample (systematically with changed start) 4 times a year with 31 90- to 140-kg (200- to 300-lb) samples, for a total of 124 samples through the year.

Suppose now we consider sampling 12 times a year ($r = 12$) instead of quarterly. At $r = 12$, s_W for aluminum is 0.0012. We note that at $n = 5$, $t = 2.132$ and

$$n_3 = [(2.132)(0.0069)]^2/[(0.002)^2(12) - (0.0012)^2(2.132)^2] = 5.2 \text{ or } 6$$

Table 2-3. Suggested Standard Deviations, s_W, for Between-Week Variation of Raw Solid Waste Components for 90- to 140-kg (200- to 300-lb) Samples

Component	Random Sampling, Quarterly	Systematic Sampling, Quarterly	Systematic/CS,[a] Quarterly	Systematic/CS, Monthly	Systematic/CS, Biweekly
Food Waste	0.0346	0.0143	0.0268	0.0135	0.0069
Garden Waste	0.0338	0.0074	0.0245	0.0183	0.0045
Total Paper	0.0586	0.0678	0.0412	0.0134	0.0036
Corrugated Paper	0.0166	0.0049	0.0137	0.0071	0.0031
Newsprint	0.0339	0.0392	0.0238	0.0077	0.0021
Plastics/Rubber/Leather	0.0088	0.0060	0.0070	0.0025	0.0024
Textiles	0.0295	0.0238	0.0242	0.0180	0.0061
Wood	0.0118	0.0035	0.0097	0.0050	0.0022
Total Organics	0.0388	0.0448	0.0312	0.0183	0.0117
Total Metals	0.0232	0.0200	0.0187	0.0137	0.0071
Ferrous Metals	0.0221	0.0177	0.0171	0.0108	0.0041
Aluminum	0.0022	0.0019	0.0020	0.0012	0.0007
Glass/Ceramics	0.0182	0.0107	0.0144	0.0128	0.0029
Ashes/Rocks/Fines	0.0234	0.0225	0.0182	0.0147	0.0105

[a]CS = changed start.

Since this is one more than we have assumed, we now try n = 6, at which point t = 2.015 and

$$n_3 = [(2.015)(0.0069)]^2/[(0.002)^2(12) - (0.0012)^2(2.015)^2] = 4.6 \text{ or } 5$$

However, a sample size of 5 is not sufficient, as shown by the previous calculation. Therefore, the required sample size is 6, and the total number of samples for the year would be 6 X 12 or 72, an improvement over 124. It pays, therefore, to examine other schedules for seasonality if the goal is to make the *total* number of samples a minimum.

It will be noted in the above example that 2 percentage points, plus or minus, on a ferrous estimate is quite a high error. If, for example, ferrous metals are in the waste at a concentration of 4%, this represents a ±50% error on the mean. Suppose we wish to reduce this error to approximately 25% of the mean, i.e., 1.0 percentage points. There is, however, a condition on Equation 2-10, i.e., $d \geqslant s_W t/\sqrt{r}$. In our example, for quarterly sampling,

$$s_W t/\sqrt{r} = (0.0171)(1.645)/\sqrt{4} = 0.014$$

so, because of the error in making ferrous measurements, the theoretical limit on precision for ferrous is 1.4 percentage points. However, we can sample more often throughout the year and change r. Let us consider sampling monthly, i.e., r = 12. Then, using the value of s_W in Table 2.3 given in the monthly systematic sampling with changed start column, we have

$$s_W t/\sqrt{r} = (0.0108)(1.645)/\sqrt{12} = 0.0051$$

and the desired precision can be achieved. Our sample size would be (by trial and error, observing that at n = 7, t = 1.943)

$$n = [(1.943)(0.0388)]^2/[(0.01)^2\, 12 - (0.0108)^2(1.943)^2] = 7.48 \text{ or } 8$$

Thus a sample size of 8, taken systematically with changed start monthly, would meet the required precision. (Note: At n = 8, t = 1.895 and n calculates to be 6.92 or 7 samples; therefore n = 8 is the correct sample size.) Would this sample size still meet our aluminum requirement of d = 0.002? The answer is "yes" because

$$s_W t/\sqrt{r} = (0.0012)(1.895)/\sqrt{12} = 0.0007$$

where 0.0012 is the s_W-value for aluminum at r = 12, and t = 1.895 is the t-value at n = 8.

In summary, the recommended procedure for sample size determination for composition is as follows:

1. Assume that quarterly (r = 4) measurements will suffice and, using the condition on Equation 2-10 for each component (i.e., $d \geqslant s_W t/\sqrt{r}$), check to see whether quarterly measurements can provide the precision required for all components to be considered.

2. If not, proceed to a consideration of additional measurements, etc., until a satisfactory sampling rate over time is found. (If monthly measurements seem too much, the s_W figures in Table 2-3 can be interpolated for values other than r = 4, 12 or 26.)
3. Now that r is selected, determine the n for all components, using Equation 2-10 and a t-value corresponding to n = infinity (e.g., at 90% confidence, this value of t = 1.645).
4. Determine the component with the highest value of n from step number 3. Refine this estimate of n by the trial and error procedure previously shown. This final n determines the weekly sample size for all components.
5. Increase r and repeat steps 3 and 4 to see whether increasing the frequency of sampling decreases the total sample size required.

If desired, one or more components may be singled out for special attention, and additional samples may be taken for this purpose. Of course, the calculation of these n-values must remain separate from the single n calculated for the other components.

2.5 STRATIFIED SAMPLING FOR COMPOSITION ESTIMATION

The composition of municipal solid waste varies with its source. For example, commercial sources would be expected to differ significantly from residential sources. Furthermore, biases may be introduced if there is great disparity in load weights (as there would be if a mixture of commercial vehicles (CV) and small, private vehicles (SPV) had access to the site), since only 90–140 kg (200–300 lb) is taken from each vehicle selected for the sample. Proportionally, a small vehicle would contribute much more to the final composition calculation.

The latter problem may be dealt with in the same manner as was discussed in Section 1.12 of Chapter 1 for quantity estimation. Equation 1-37 would be used to determine sampling rates between the two vehicle types, and the following modifications of Equations 1-38 and 1-40 would be used for the estimation of component fraction:

$$C_0 = \sum_{i=1}^{2} (N_i/N)C_i \tag{2-11}$$

$$\mathrm{var}(C_0) = \sum_{i=1}^{2} (N_i/N)^2\, \mathrm{var}(C_i) \tag{2-12}$$

where C_0 is the overall or combined estimate of fraction composition, C_i is the estimate of fraction composition for SPV type (i = 1) or CV type (i = 2) obtained from Equation 2-7, and $\mathrm{var}(C_i)$ is the variance obtained from Equation 2-8. The N_i and N have the same meanings as in Equations 1-38 and 1-40.

With regard to the difference between the residential and commercial sources within the CV types, these two generally do not exhibit great disparity in load weights. If they did, however, then the procedures just outlined would be applicable in that case also.

2.6 SAMPLE WEIGHT AND SAMPLE SIZE FOR PROCESS STREAMS

The 90- to 140-kg (200- to 300-lb) sample weight used in the previous sections of this chapter is appropriate for raw refuse only. Clearly, the outputs of intermediate processing steps in a resource recovery system (such as shredding and air classification) do not require sample weights anywhere near this magnitude. The determination of an optimal sample weight is very important in the case of raw refuse because separating raw refuse is expensive, and the cost is proportional to the sample weight. With regard to process streams, on the other hand, optimality of sample weight is far less important. Process streams are much more homogeneous than is raw refuse. Further, hand separation is generally not involved. Finally, many of the measurements of process streams we are interested in (percent ashes, Btu content, etc.) are fairly independent of sample weight. Frequently, in fact, a process sample is ground (to reduce variability) and a standard amount is selected for the test (or used for a composite) with the remainder being discarded. For this reason, there is no reason to approach the problem of optimal weight for process stream samples with overly sophisticated techniques.

Optimal sample weight is related to the particle size of the material sampled, but the relationship certainly is not linear. Other investigators [9] have found that process stream particle size distributions are adequately described by functions based on the exponential distribution, the Rosin–Rammler equation, for example. Further, the author has observed that the composition distribution of raw refuse can be described very satisfactorily by functions involving exponential terms. For example, in the computer sampling experiments described in Section 2.3, products of exponential distributions were used to generate values for wood and textiles, and sums of exponential distributions were used to generate values for the other components. These functions duplicate real-world data very well.

For these reasons, the author recommends the following model based on the exponential function to determine optimal sample weight for process streams:

$$Y = Xe^{aX} \tag{2-13}$$

where Y is the optimal sample weight in pounds and X is the characteristic particle size of the material to be sampled, in inches. The boundary condition for Equation 2-13 is that at Y = 115 kg (250 lb) (the median of 90 and 140 kg or 200 and 300 lb), the characteristic particle size (i.e., the screen size at which 63.2% of the material passes through) is 46 cm (18 in.). The value of a that satisfies the condition is 0.146. Thus,

$$Y = Xe^{0.146X} \tag{2-14}$$

To illustrate use of this equation, we note that the output of the Appelton West shredder was found to have a characteristic size of 5.6 cm (2.2 in.). (This is based on information in Table 10 of a report by Savage and Shiflett [10].) To sample the output from this shredder we would need to take samples of weight

$$Y = 2.2\ e^{0.146(2.2)} = 3.0 \text{ lb}$$

On the other hand, the characteristic size of the Ames secondary shredder output was found to have an average characteristic size of 1.35 cm (0.53 in.). (This is based on information in Table 10 of a report by Savage and Shiflett [10].) The sample weight for sampling from this shredder would be

$$Y = 0.53\ e^{0.146(0.53)} = 0.57 \text{ lb}$$

With regard to the number of samples of size Y required, Equation 2-5 is applicable to process streams as well as to raw refuse. For example, the median characteristic size of the light fraction of the Tacoma, Washington air classifier is 1 in. This information comes from Table 4a of a report by Simister [11]. The average percentage of ash in the light fraction is 13.3%, with a standard deviation, s, of 2.02%. This information is taken from Table A-1 of a report by Simister [11]. From Equation 2-14, the sample weight is

$$Y = 1X^{0.146} = 1.16 \text{ lb}$$

Suppose we wish to estimate (at 90% confidence) the average percentage of ashes to within 20% of its mean. Then d = 0.20 (13.3) = 2.66%. Using Equation 2-5 and a t-value of 2.353 at n = 4, we have:

$$n = \left[\frac{(2.353)(2.02)}{2.66}\right]^2 = 3.2 \text{ or } 4 \text{ samples}$$

The above calculation depended, of course, on a prior estimate of the standard deviation. Unfortunately, not enough is known about resource recovery process streams to provide such estimates for all measurements of interest. One approach may be helpful, however. The method is as follows [12]:

1. Make a guess for the value of s.
2. From this, use Equation 2-5 to determine n, the size of the sample. Take a first sample whose size is 0.5n–0.7n.

3. This first sample, n_1, will provide an estimate of the population standard deviation.
4. Using the value of standard deviation found in Step 3, recompute n in Equation 2-5. The recomputed value of n will be the total required sample size for the first and second samples combined.

As an example, suppose in our previous example we did not know s but estimated it as 3%. Then, from Equation 2-5 (using a t-value of 2.015 at n = 6),

$$n = \left[\frac{(2.015)(3)}{2.66}\right]^2 = 5.2 \text{ or } 6 \text{ samples}$$

We now let $n_1 = 0.5n = 3$. Suppose we take a sample of 6 and set 3 of them aside, and suppose further that the s calculated from the 3 samples actually measured was 2.02. Then from Equation 2-5 (using a t-value of 2.353 at n = 4),

$$n = \left[\frac{(2.353)(2.02)}{2.66}\right]^2 = 3.2 \text{ or } 4 \text{ samples}$$

We need, therefore, to measure one more of the three samples set aside.

2.7 SUMMARY

Compared with the estimation of quantity, the estimation of municipal solid waste composition is much more complex. Unfortunately, sampling for composition is bound to be both more expensive and less precise than for its quantity counterpart. This is primarily because separation for composition is done by hand and the decision to classify an item of solid waste is judgmental at times. Bearing in mind this fundamental problem of accuracy in classification, the techniques presented in this chapter will allow the planner or designer of a municipal solid waste resource recovery system to make useful estimates.

As in the case of quantity estimation, the general problem of seasonality has received little or no attention in the past. The matter has been discussed in great detail here, however, and the methodology developed poses no great problems except for the usual one of cost. Information is rarely free, however, at least information of the useful kind.

Also, as in Chapter 1, a methodology for the estimation of sample size was developed using actual data and statistics generated from computer sampling experiments. We recognize once again that the data bases used neither reflect all real-world situations nor do they necessarily duplicate the planner's or designer's specific situation. Rule-of-thumb estimates of critical variances are invaluable, however, when site-specific data are not available. Also, no matter what the size of the sample actually taken, the equations presented in this chapter for estimating the average fraction of components (and the variances of these averages) remain valid at all times.

Finally, although there are few hard data available on the composition of process streams, suggestions have been made for sample weight and sample size in these instances that are based on both theoretical and empirical considerations.

REFERENCES

1. Kish, L. "Cluster Sampling," in *Survey Sampling* (New York: John Wiley & Sons, Inc., 1965), p. 161.
2. Klee, A. J., and D. Carruth. "Sample Weights in Solid Waste Composition Studies," *J. San. Eng. Div., ASCE* 96(SA4):945-954 (1970).
3. Tortora, R. D. "A Note on Sample Size Estimation for Multinomial Populations," *Am. Statist.* 32(3):100-102 (1978).
4. Cochran, W. G. "The Estimation of Sample Size," in *Sampling Techniques*, 2nd ed. (New York: John Wiley & Sons, Inc., 1963), pp. 74,75.
5. Weast, R. C., Ed. "Ten-Place Values for Inverse Sines and Tangents," in *Handbook of Tables for Mathematics* (Cleveland, OH: Chemical Rubber Co., 1967), pp. 448-457.
6. Mace, A. E. "Estimation Problems," in *Sample-Size Determination* (New York: Van Nostrand Reinhold Co., 1964), pp. 35-37.
7. Britton, P. W. "Improving Manual Solid Waste Separation Studies," *J. San. Eng. Div., ASCE* 98(SA5):717-730 (1972).
8. Peter Middleton and Associates, Ltd. *A Municipal Solid Waste Management Study*, The Pollution Probe Foundation, Toronto, Ontario (1974), pp. 1-86.
9. Trezek, G. "Significance of Size Reduction in Solid Waste Management," EPA-600/2-77-131, U.S. Environmental Protection Agency, Cincinnati, OH (1977).
10. Savage, G. M., and G. R. Shiflett. "Processing Equipment for Resource Recovery Systems: III–Field Test Evaluation of Shredders," Contract No. 68-03-2589, U.S. Environmental Protection Agency, Cal Recovery System, Richmond, CA (1979).
11. Simister, B. W. "Test Report: Comparative Study of Air Classifiers, Tacoma, Washington," Contract No. 68-03-2730, U.S. Environmental Protection Agency, Midwest Research Institute, Kansas City, MO (1979).

CHAPTER 3

UNCERTAINTY ANALYSIS, DECISION TREES AND MULTIATTRIBUTE DECISION-MAKING

3.1 INTRODUCTION TO UNCERTAINTY ANALYSIS

In its simplest form, uncertainty analysis consists of: (1) estimating the probability distribution of each factor affecting a resource recovery calculation; (2) for each factor, selecting a value at random from its distribution and combining these values to determine a particular outcome for the calculation; and (3) repeating step 2 a number of times so that a number of outcomes is obtained. In short, instead of presenting the decision-maker with a single number, a distribution of values is provided. There are at least three advantages to this method:

1. While the single number is virtually certain to be incorrect, when using uncertainty analysis there is a good chance that the correct value lies within the range of the estimated distribution.
2. Uncertainty necessarily becomes more salient to the decision-maker with this approach, i.e., the absoluteness of the single number is absent.
3. Uncertainty becomes more understandable to the decision-maker inasmuch as both the shape and the range of the uncertainty distribution are presented clearly.

A simple example will illustrate the technique. Figure 3-1 shows a traditional cost calculation flow diagram that involves six uncertain quantities: capital and direct operation and maintenance (O&M) costs, plus four factors used in the calculations—for amount of funds during construction (AFDC), working capital (WC), administrative overhead (AO), and real estate taxes and insurance (RETI). In Figure 3-1 the four factors are indicated as constants: 0.05, 0.0833, 0.2 and 0.02, respectively. Ordinarily, both capital and direct O&M costs would also be introduced as constants into the calculation. For example, we suppose the following:

Capital Costs	= \$513,126
Direct O&M Costs	= \$763,380
Yearly Interest on Capital	= 10%
Annual Inflation Rate	= 7%
Discount Factor	= 10%
Plant Life	= 5 years

Then the following costs are computed using Figure 3-1:

	Present Value Annualized Costs (PVAC)
Year 0	\$1087K
Year 1	\$1046K
Year 2	\$1008K
Year 3	\$ 967K
Year 4	\$ 927K
Total	\$5035K

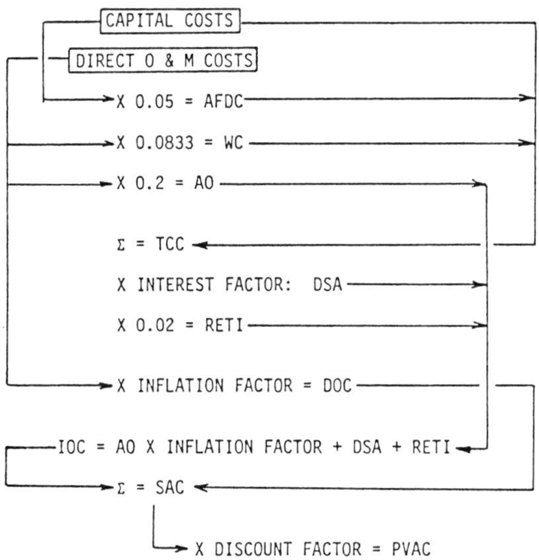

KEY

AFDC = Amount of Funds During Construction
WC = Working Capital
AO = Administrative Overhead
TCC = Total Capital Costs
DSA = Debt Service & Amortization
RETI = Real Estate Taxes & Insurance
DOC = Direct Operating Costs
IOC = Indirect Operating Costs
SAC = Sum Annualized Costs
PVAC = Present Value Annualized Costs

INTEREST FACTOR = $i(1 + i)^N/[(1 + i)^N - 1]$
where i is yearly interest on capital and N is total years of plant life.

INFLATION FACTOR = $(1 + j)^n$
where j is annual inflation rate and n is nth year.

DISCOUNT FACTOR = $(1 + k)^n$
where k is annual discount rate and n is nth year.

Figure 3-1. Cost calculation flow diagram.

To do an uncertainty analysis, we once again use the flow diagram of Figure 3-1, but this time the capital costs, direct O&M costs, and AFDC, WC, AO and RETI factors are treated as stochastic variables. To illustrate the technique, we assume that these six factors are normally distributed, with their standard deviations equal to 20% of their respective means. (See Chapter 4 for techniques to obtain probability distributions of an ad hoc nature.) The calculations of the flow diagram are completed as before, this time using values selected at random from these six distributions. The process is repeated 1000 times, clearly a task for a digital computer. The simulation language preferred by the author is GPSS, primarily because of its programming simplicity, but it should be remembered that great precision is *not* required. Because of the uncertainty of many of the inputs, it makes no sense to carry calculations beyond three or four significant figures. Accordingly, FORTRAN programs are not required. The arithmetic available to the GPSS user admittedly is limited, but it more than makes up for this by the ease it affords in handling ad hoc stochastic distributions. Such a GPSS program is shown in Figure 3-2, and the output PVAC are plotted in Figure 3-3, along with the deterministic PVAC values calculated previously. The cumulative probability distribution of the total PVAC is shown in Table 3-1, which shows an approximately 50% probability that the total PVAC for this system lie between \$4850K and \$5250K.

For those readers desiring to run this particular simulation using different parameters for the distributions, the SAVEVALUE locations are initialized as follows:

X10 - Mean of capital cost distribution
X11 - Mean of direct O&M cost distribution
X12 - Mean of AFDC factor distribution × 10,000
X13 - Mean of WC factor distribution × 10,000
X14 - Mean of AO factor distribution × 1000
X15 - Mean of RETI factor distribution × 10,000
X16 - SD of capital cost distribution
X17 - SD of direct O&M cost distribution
X18 - SD of AFDC factor distribution × 10,000
X19 - SD of WC factor distribution × 10,000
X20 - SD of AO factor distribution × 1000
X21 - SD of RETI factor distribution × 10,000
X50 - Interest factor × 10,000
X51 - (1 + j) × 100
X52 - (1 + k) × 100
X53 - N, total number of years

Finally, it should be mentioned that the determination of capital and direct O&M costs themselves involves computations that include uncertain factors. Therefore, these two quantities could also have been subjected to uncertainty analysis in their own right.

```
          SIMULATE                              UNCERTAINTY ANALYSIS
          INITIAL     X10,K513126
          INITIAL     X11,K763380
          INITIAL     X12,K500
          INITIAL     X13,K833
          INITIAL     X14,K200
          INITIAL     X15,K200
          INITIAL     X16,K34208
          INITIAL     X17,K50892
          INITIAL     X18,K33
          INITIAL     X19,K53
          INITIAL     X20,K13
          INITIAL     X21,K13
          INITIAL     X50,K2638
          INITIAL     X51,K107
          INITIAL     X52,K110
          INITIAL     X53,K5
          INITIAL     X54,K100                  CONSTANT
1         FUNCTION    RN2,C25
0,-5/.00003,-4/.00135,-3/.00621,-2.5/.02275,-2
.06681,-1.5/.11507,-1.2/.15866,-1/.21186,-.8/.27425,-.6
.34458,-.4/.42074,-.2/.5,0/.57926,.2/.65542,.4
.72575,.6/.78814,.8/.84134,1/.88493,1.2/.93319,1.5
.97725,2/.99379,2.5/.99865,3/.99997,4/1,5
1         FVARIABLE   FN1*X16+X10
2         FVARIABLE   FN1*X17+X11
3         FVARIABLE   ((FN1*X18+X12)*X100)/K10000
4         FVARIABLE   ((FN1*X19+X13)*X101)/K10000
5         FVARIABLE   ((FN1*X20+X14)*X101)/K10000
6         FVARIABLE   (X101*V20)/K10000
7         FVARIABLE   X100+X102+X103
8         FVARIABLE   (X50*X106)/K10000
9         FVARIABLE   ((FN1*X21+X15)*X106)/K10000
10        FVARIABLE   (X104*V20)/K10000
11        FVARIABLE   X109+X107+X108
12        FVARIABLE   X110+X105
13        FVARIABLE   ((X111*V21)/K10000)/K1000
20        FVARIABLE   X60*61
21        FVARIABLE   X62*X63
22        FVARIABLE   (X101*V20)/K10000
23        FVARIABLE   X107+X108+(X104*V20)/K10000
24        FVARIABLE   ((X200+X201)*K10)/V21
25        FVARIABLE   V20/K100
26        FVARIABLE   V21/K100
```

Figure 3-2. GPSS simulation program

3.2 INTRODUCTION TO DECISION TREES

When a decision-maker (or group of decision-makers) must select one act from a set of possible actions, the basic optimal procedure is to examine the consequence of each act, determine the probabilities of occurrence of the events that affect those consequences and, following this, to determine the act having the greatest expected utility to the decision-maker(s). For example, suppose the directors of a resource recovery plant have the option of adding a certain glass recovery subsystem to the existing system. The subsystem is of a radical new design and there is considerable uncertainty that

```
1       TABLE      X112,800,20,100
2       TABLE      X112,800,20,100
3       TABLE      X112,780,20,100
4       TABLE      X112,780,20,100
5       TABLE      X112,740,20,100
6       TABLE      X210,4000,100,100
        GENERATE   ,,,1
LOOP3   SAVEVALUE  60,X54
        SAVEVALUE  61,X54
        SAVEVALUE  62,X54
        SAVEVALUE  63,X54
        SAVEVALUE  100,V1
        SAVEVALUE  101,V2
        SAVEVALUE  102,V3
        SAVEVALUE  103,V4
        SAVEVALUE  104,V5
        SAVEVALUE  105,V6
        SAVEVALUE  106,V7
        SAVEVALUE  107,V8
        SAVEVALUE  108,V9
        SAVEVALUE  109,V10
        SAVEVALUE  110,V11
        SAVEVALUE  111,V12
        SAVEVALUE  112,V13
LOOP2   ASSIGN     1+,K1
        SAVEVALUE  210+,X112
        TABULATE   *1
        TEST E     P1,X53,LOOP1
        TABULATE   6
        SAVEVALUE  210,K0
        ASSIGN     1,K0
        SPLIT      1,LOOP3
        TERMINATE  1
LOOP1   SAVEVALUE  60,V25
        SAVEVALUE  62,V26
        SAVEVALUE  61,X51
        SAVEVALUE  63,X52
        SAVEVALUE  200,V22
        SAVEVALUE  201,V23
        SAVEVALUE  112,V24
        TRANSFER   ,LOOP2
        START      1000
        END
```

for example of Section 3.1.

it will work. If it does work, the discounted (for time) net profit over the life of the plant is \$300,000; if the system does not work properly, the discounted net losses will be \$100,000. (The net profits and losses include amortization and operating costs.) The probability of success is estimated as 50%. Accordingly, the decision picture would look like that shown in Figure 3-4 where a decision tree is pictured. The tree is made up of nodes and branches, each node being of two basic types: squares, denoting decisions; and circles, denoting uncertain or chance events. The branches emanating from decision nodes represent alternatives available to the decision-maker; branches from chance event nodes represent alternative

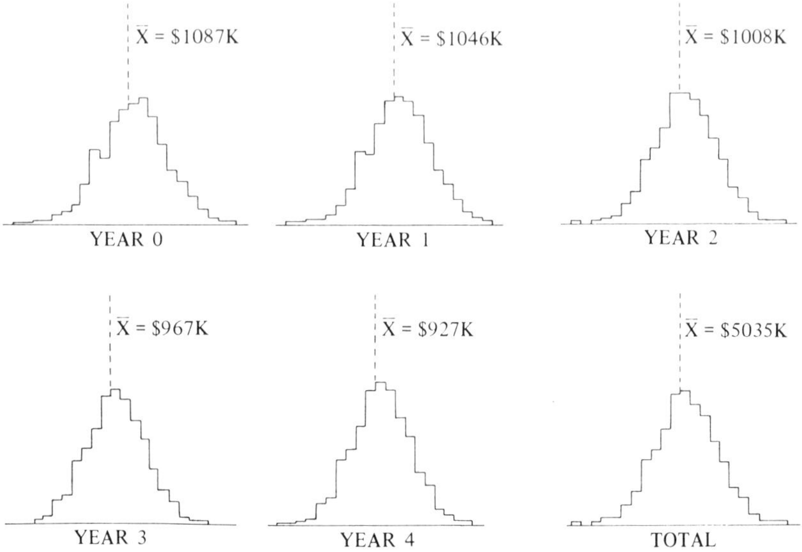

Figure 3-3. Output distributions for example of Section 3.1.

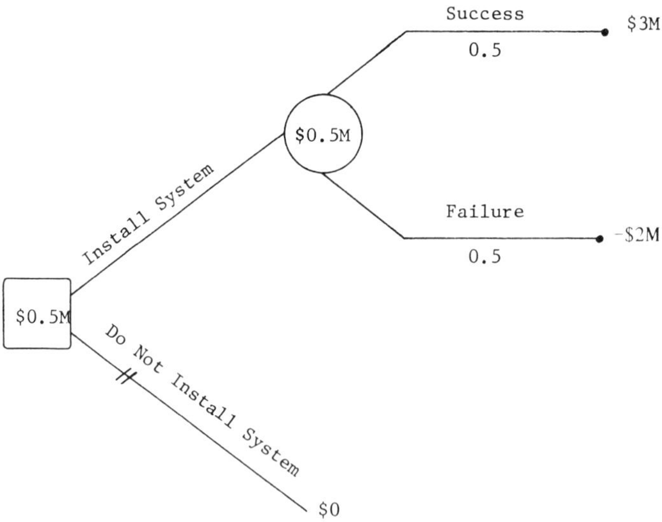

Figure 3-4. Simple decision problem of Section 3.2.

Table 3-1. Cumulative Probability Distribution for Total Present Value Annualized Costs

Upper Limit ($)	Cumulative Probability (%)
4000K	0.1
4100	0.2
4200	0.4
4300	1.0
4400	1.7
4500	3.7
4600	6.9
4700	13.2
4800	20.8
4900	30.1
5000	43.1
5100	56.2
5200	68.5
5300	79.1
5400	87.1
5500	93.2
5600	96.5
5700	98.9
5800	99.8
5900	99.8
6000	99.8
6100	100.0

outcomes of these events. Finally, the consequences are placed at the ends of the branches.

In this example, the expected value of the single chance node is $500,000, i.e., 0.5(3,000,000) + 0.5(−2,000,000). Thus, the expected value of the act "Install" is $500,000 and of "Do not install" is $0. The optimal act, therefore, is "Install." This type of problem is well known and needs no further discussion.

There is another type of problem that is quite different, however, and this occurs when the problem involves a *sequence* of decisions, rather than only a single one [1]. For example, we might modify our original problem in the following manner. Suppose the directors have another option, i.e., to wait one year for the designers to "debug" the system and make further improvements. This would increase the system's success probability to 0.7. (There would also be small increases in costs and small decreases in net profits because the system would be discounted an additional year farther out in the time scale.) This new situation is represented in Figure 3-5. The

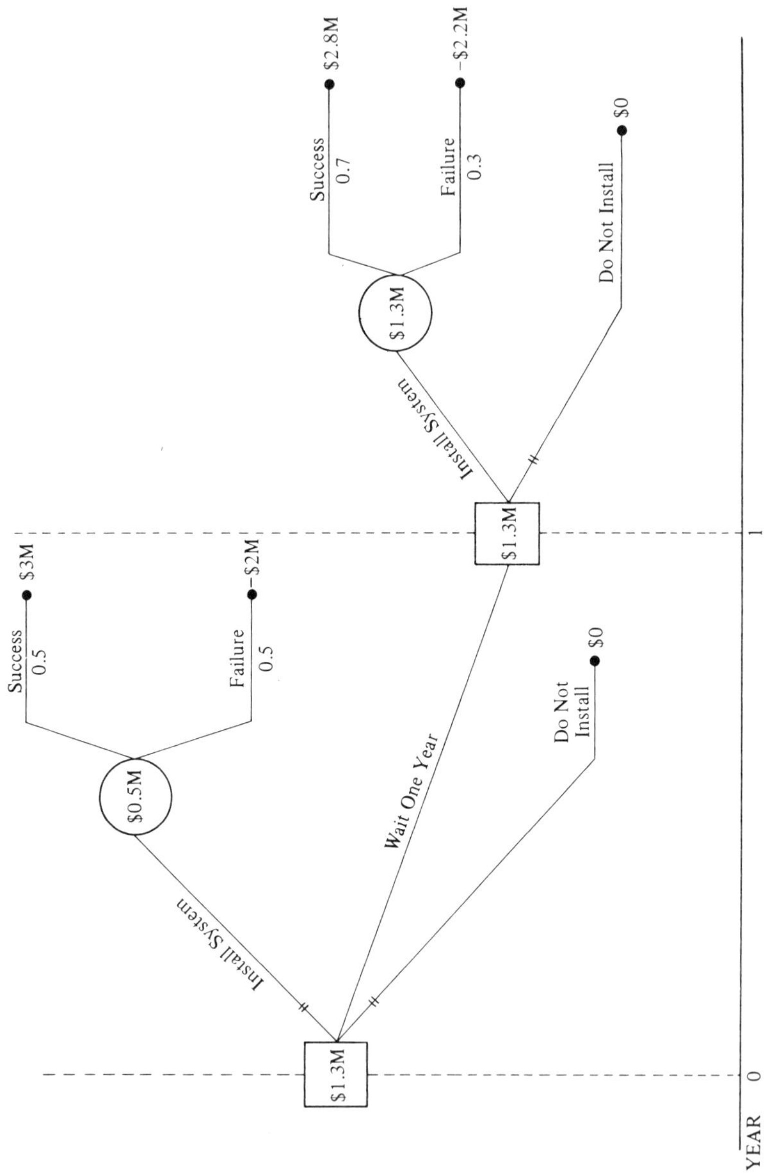

Figure 3-5. Simple decision problem of Section 3.2 involving time.

solution is obtained by the application of a "rollback" principle in which we consider only optimal returns from each decision node, starting from the ends of the tree. This approach works because if we assume that we are on the optimal path, decisions that have been made in the past are irrelevant and need not be taken into consideration at that point in time.

For example, at the lower right-hand corner of the decision tree pictured in Figure 3-5, the expected value of "Install" is $1,300,000, while the expected value of "Do not install" is $0. Although we do not know whether we actually are on the optimal path, *if* we were, then clearly the optimal decision at that point in time would be to "Install." In this manner, we work back to time zero. Since, working backwards, all decisions at a future time are made optimally, the decision at time zero must be optimal. A more complex decision tree, with solution, is shown in Figure 3-6. (The two cross-bars present on some decision branches in these decision trees indicate rejected decisions.)

It should be noted that the concept of time in a decision tree is relative. In Figure 3-5, a year must elapse before a second look at the decision "Install/Not Install" can be made, assuming the system was not installed at time zero. In Figure 3-6, the times t_0, t_1 and t_2 may be separated by weeks, months or years, etc., or the decisions at these times may be almost simultaneous. The point is that the A-pair decision must be made before the B-pair, and the B-pair before the C-pair. It is the sequence that is important not the time interval itself (except that all values on the decision tree should be converted to present values).

Another important point is that decision trees *do not deal with future decisions*. It deals with the futurity of present decisions. In Figure 3-6, for example, our solution is to adopt alternative A_2; nothing has been decided about the B-pair or C-pair decisions. For one thing, at time t_1, the probabilities or consequences may change significantly [2].

3.3 UNCERTAINTY IN DECISION TREES

The simplest form of decision tree is that in which (1) there are a finite number of events; (2) there are a finite number of alternatives or decisions; and (3) the values of the outcomes are deterministic (i.e., single-valued) in nature. There are times, however, when these are not realistic assumptions. We have already dealt with the concept of uncertainty in a nondecision tree context in Section 3.1. Let us consider the following example. It is proposed to add a glass recovery system to an existing resource recovery plant. There are two options:

1. Build a small plant (cost = $300,000) of 3000 ton/yr capacity.
2. Build a large plant (cost = $600,000) of 10,000 ton/yr capacity.

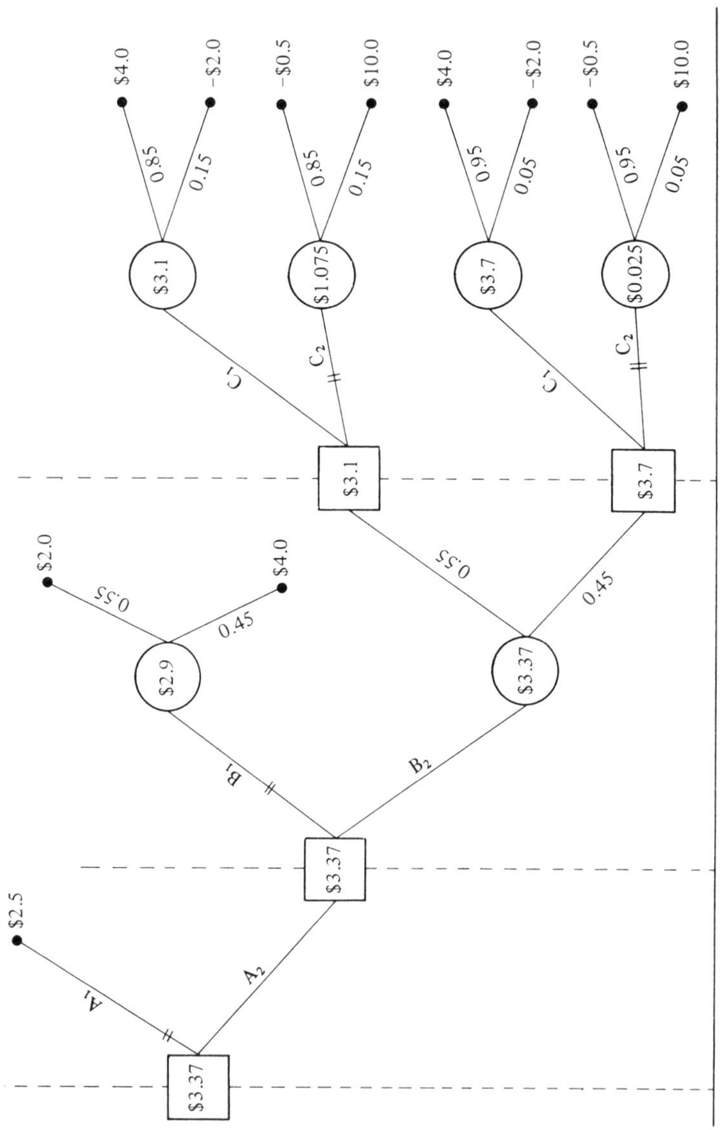

Figure 3-6. Complex decision tree showing application of the "rollback" principle.

If a small plant is built, however, it is possible to expand the plant to a capacity of 10,000 ton/yr at an additional cost of \$450,000, should the demand justify it. Demand is estimated to vary between 0 and 10,000 ton/yr, and the present value of each ton demanded is \$200. The basic equation used to calculate net profit is simply

$$\text{Net Profit} = (\text{Demand})(\text{Present Value}) - \text{Plant Cost} \quad (3\text{-}1)$$

Assuming certain demands, the problem may be represented as the decision tree shown in Figure 3-7. The "justifiable" demand for plant expansion is assumed to be 4000 ton/yr or more, and the actual average demands used in the decision tree may be back-calculated from the net profits given. For example, on the "Moderate Demand, Do Not Expand" branch, the demand used was 2400 ton/yr because, using Equation 3-1, 180,000 = 2400(200) − 300,000. In any event, following the usual rollback principle, the solution to the decision tree is "Build Large Plant," since it has the greatest expected value. But is it? The careful reader will note that the "optimal" solution involves a 20% chance of a \$200,000 net loss; the other alternative involves no losses at all; further, on the average it returns only 6% less than the "optimal" solution. Clearly, the traditional decision tree does not handle uncertainty very well.

3.4 THE STOCHASTIC DECISION TREE

The solution to this problem is the stochastic decision tree [3], which utilizes the principle identical to the uncertainty analysis of Section 3.1, i.e., a simulation of the probable combinations in the tree to produce a distribution of outcomes for any given alternative. First, let us assume a specific demand distribution, i.e., the one shown in Table 3-2. (This particular distribution is unimodal, skewed to the right.) The decision tree for this problem is shown in Figure 3-8. Distributions A, B, C and D are the same, i.e., the demand distribution of Table 3-2. Only the shaded portions, however, are used on their respective branches. On branch A, for example, only demands of less than 4000 ton/yr are realized; on branches B and C only demands of greater than or equal to 4000 ton/yr are realized; on branch D, all demands from 0–10,000 ton/yr are realized. Distributions E, F and G are net profit distributions derived from the demand distributions using Equation 3-1.

It is clear that there are three alternatives available to the decision-maker:

1. Build large plant.
2. Build small plant.
3. Build small plant and enlarge if demand exceeds 4000 ton/yr.

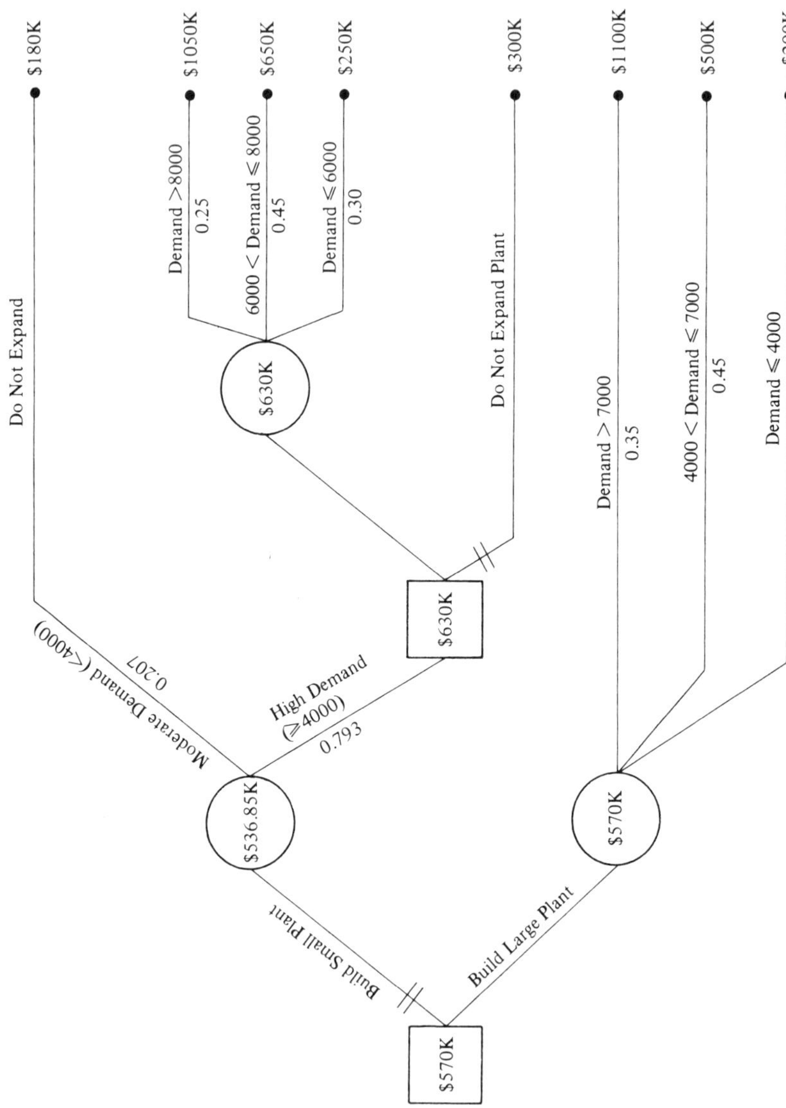

Figure 3-7. Example of Section 3.3.

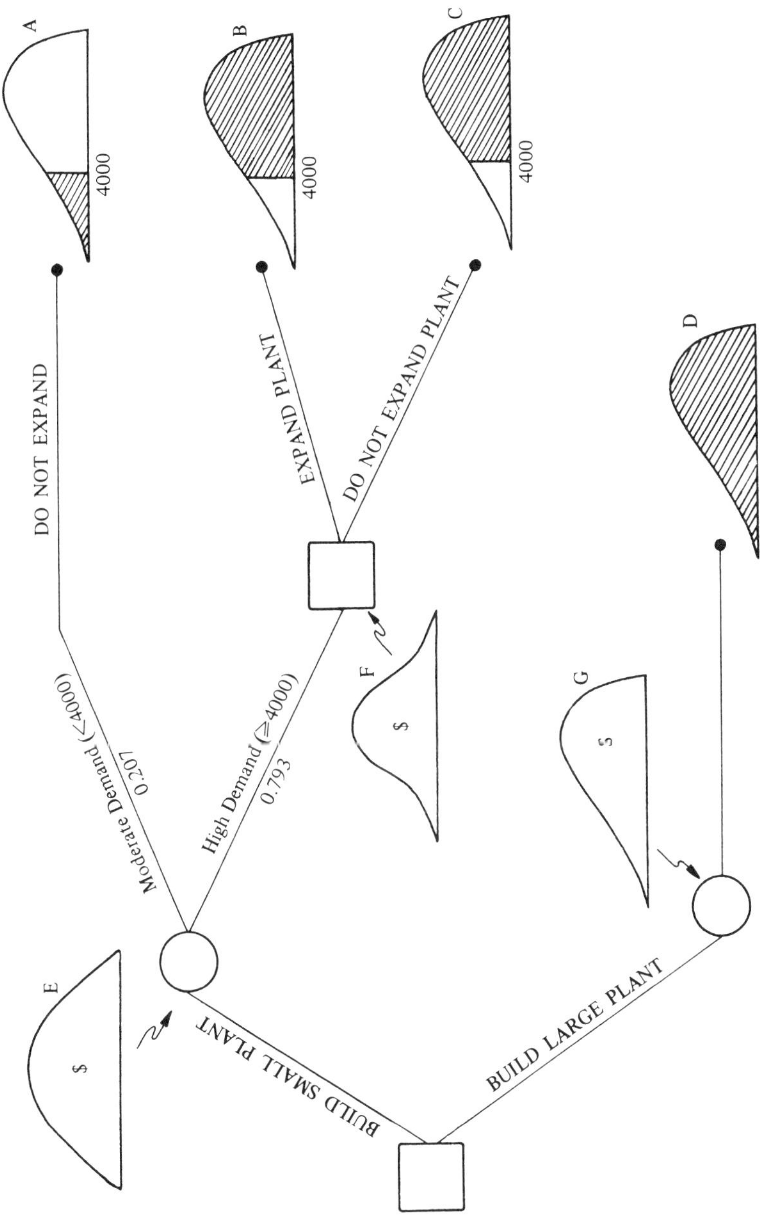

Figure 3-8. Example of Section 3.4.

Table 3-2. Cumulative Demand Distribution for Sample Problem of Section 3.4

D = Demand (ton/yr)	Probability (Demand ⩽ D)
1,000	0.013
2,000	0.051
3,000	0.115
4,000	0.207
5,000	0.327
6,000	0.471
7,000	0.634
8,000	0.804
9,000	0.946
10,000	1.000

The distribution of outcomes for these three alternatives are obtained via simulation. As in Section 3.1, GPSS is the recommended simulation language for its ease of implementation. These three alternatives are simulated in the GPSS programs shown in Figures 3-9 and 3-10. These programs incorporate a further complication, i.e., the present value factor in Equation 3-1 *is now made stochastic as well* (normally distributed with a mean of 200 and a standard deviation of 30). This shows the ability of GPSS to deal, in a very simple and straightforward manner, even with complex decision tree problems. The outputs are the usual GPSS distribution tables, which are shown in the graphics of Figure 3-11.

The results in Figure 3-11 no doubt will be surprising to some. There is a large peak in the net profit distribution of alternative 2 (build small plant) at net profit = \$300. This is because the small plant is subjected to demands that exceed its capacity (>3000 ton/yr) approximately 89% of the time (see Table 3.2). There is a similar (but not as severe) peak for alternative 3 because even for this alternative the capacity of the small plant is exceeded approximately 44% of the time (i.e., from Table 3.2, 0.44 = (0.207 − 0.115)/0.207). Figure 3-11 clearly eliminates alternative 2 as a solution, but the choice between alternatives 1 and 3 is not as clear. The expected value of the former exceeds that of the latter (by about 9%), but there is more than four times the chance of a loss with alternative 1. From the GPSS output tables (not shown here), *if* a loss is sustained, the average value of the loss would be \$110.4K with alternative 3 and \$218.4K with alternative 1. The decision-maker might well opt for alternative 3 with its lower risk than for alternative 1 with its somewhat higher average return.

It will be noted that the rollback principle is not applicable in the solution

```
      SIMULATE   DECISION TREE ALTERNATIVES 1 AND 2
PRESV EQU        1,Z
DMAND EQU        2,Z
PRESV FUNCTION   RN2,C25
0,-5/.00003,-4/.00135,-3/.00621,-2.5/.02275,-2
.06681,-1.5/.11507,-1.2/.15866,-1/.21186,-.8/.27425,-.6
.34458,-.4/.42074,-.2/.5,0/.57926,.2/.65542,.4
.72575,.6/.78814,.8/.84134,1/.88493,1.2/.93319,1.5
.97725,2/.99379,2.5/.99865,3/.99997,4/1,5
DMAND FUNCTION   RN1,C11
0.0/.013,1000/.051,2000/.115,3000/.207,4000/.327,5000/.471,6000
.634,7000/.804,8000/.946,9000/1,10001
ALT2  TABLE      X50,-300,50,200
ALT1  TABLE      X53,-600,50,200
NPV   FVARIABLE  (F$PRESV*K30)+K200
AVAR  FVARIABLE  (X10*V$NPV-K300000)/K1000
DVAR  FVARIABLE  (X10*V$NPV-K600000)/K1000
      RMULT      ,71
      GENERATE   ,,,1
LOOP1 SAVEVALUE  10,FN$DMAND
      TEST LE    X10,K3000,LOOP2
LOOP3 SAVEVALUE  50,V$AVAR
      TABULATE   ALT2
      SAVEVALUE  10,FN$DMAND
      SAVEVALUE  53,V$DVAR
      TABLUATE   ALT1
      SPLIT      1,LOOP1
      TERMINATE  1
LOOP2 SAVEVALUE  10,K3000
      TRANSFER   ,LOOP3
      START      1000
      END
```

Figure 3-9. GPSS simulation program for alternatives 1 and 2 of problem of Section 3.4.

```
      SIMULATE DECISION TREE ALTERNATIVE 3
PRESV EQU        1,Z
DMAND EQU        2,Z
PRESV FUNCTION   RN2,C25
0,-5/.00003,-4/.00135,-3/.00621,-2.5/.02275,-2
.06681,-1.5/.11507,-1.2/.15866,-1/.21186,-.8/.27425,-.6
.34458,-.4/.42074,-.2/.5,0/.57926,.2/.65542,.4
.72575,.6/.78814,.8/.84134,1/.88493,1.2/.93319,1.5
.97725,2/.99379,2.5/.99865,3/.99997,4/1,5
DMAND FUNCTION   RN1,C11
0,0/.013,1000/.051,2000/.115,3000/.207,4000/.327,5000/.471,6000
.634,7000/.804,8000/.946,9000/1,10001
ALT3B TABLE      X54,-300,50,200
NPV   FVARIABLE  (F$PRESV*K30)+K200
AVAR  FVARIABLE  (X10*V$NPV-K300000)/K1000
BVAR  FVARIABLE  (X10*V$NPV-K750000)/K1000
      RMULT      ,71
      GENERATE   ,,,1
LOOP6 TRANSFER   .200,,LOOP3
LOOP4 SAVEVALUE  10,FN$DMAND
      TEST G     X10,K4000,LOOP4
      SAVEVALUE  54,V$BVAR
      TABLULATE  ALT3
      TRANSFER   .LOOP5
LOOP3 SAVEVALUE  10,FN$DMAND
      TEST LE    X10,K4000,LOOP3
      TEST G     X10,K3000,LOOP7
      SAVEVALUE  10,K3000
LOOP7 SAVEVALUE  54,V$AVAR
      TABLUATE   ALT3
LOOP5 SPLIT      1,LOOP6
      TERMINATE  1
      START      1000
      END
```

Figure 3-10. GPSS simulation program for alternative 3 of problem of Section 3.4.

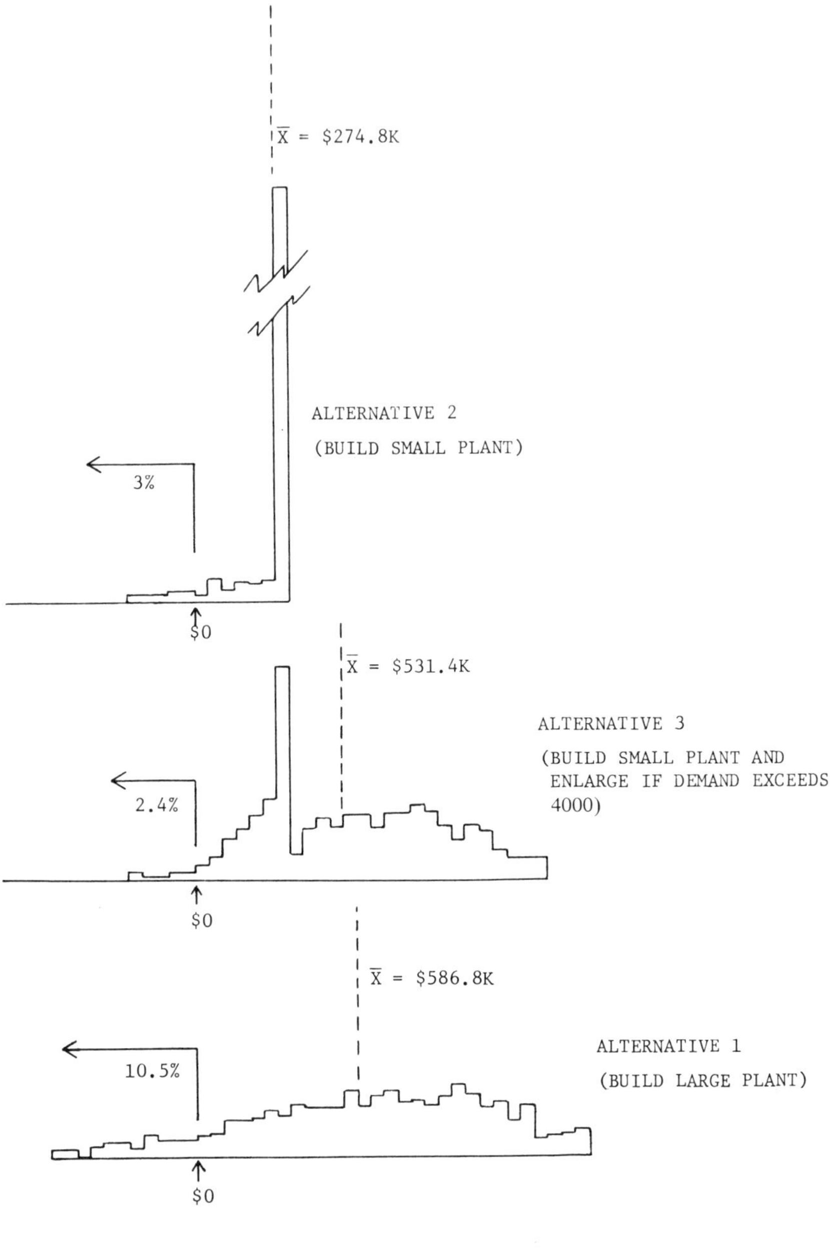

Figure 3-11. Net profit distributions for example of Section 3.4.

to stochastic decision trees because the criterion of expected value is no longer the sole criterion. Since the stochastic decision tree is based on simulation, it is not necessary to add as many branches at chance event nodes. In fact, as is shown in Figure 3-8, it is frequently possible to reduce the number of branches to one. In effect, the chance event node is eliminated, resulting in a drastic pruning of the tree. Thus, it is often possible to evaluate by complete enumeration all of the possible paths through the decision tree, as was done here.

3.5 NONFINITE NUMBER OF ALTERNATIVES

The example problem of Sections 3.3 and 3.4 involves a finite number of alternatives (build large plant, build small plant, build small plant and enlarge if demand exceeds 4000 ton/yr) only because an arbitrary cutoff of 4000 ton/yr was made on the "build small plant" branch. In theory, there are an infinite number of alternatives possible as this decision point varies continuously from 0 through 10,000 ton/yr of demand. Practically, of course, differences in demand of less than some stated value (perhaps 200 ton/yr in this example) would not be considered important by the decision-maker, so that the number of different demands recognized might be reduced from infinite to something like 50. With 50 branches at this point, however, the tree becomes a bushy mess.

The simulation approach of Section 3.4, however, makes the problem tractable once again. The cutoff point (which was arbitrarily set at 4000 ton/yr in our example) is made a parameter, X, and the alternative (build small plant and enlarge if demand exceeds X ton/yr) then is simulated for varying values of X. The GPSS program to do this for our sample problem is shown in Figure 3-12 and represents but a slight modification of the program presented in Figure 3-10. Nine simulation runs were made, representing parameter values ranging from 3000–6000 ton/yr, the results being shown in Figure 3-13 where the means of the net profit distributions are plotted. The curve is rather flat between 3800 and 5300 ton/yr, but a maximum value ($540K) is obtained at X = 4400 ton/yr. This further enhances the desirability of this alternative.

3.6 MULTIATTRIBUTE DECISION-MAKING

Although the multiplicity of objectives in many areas of decision-making can be readily recognized and accepted, multiple-goal decision-making has traditionally been approximated by single, unchanging and technically manageable criteria. In the example problems of this chapter, for example,

```
        SIMULATE           DECISION TREE, STOCHASTIC EVENT
        INITIAL    X100,4200
 PRESV  EQU        1,Z
 DMAND  EQU        2,Z
 PRESV  FUNCTION   RN2,C25
0,-5/.00003,-4/.00135,-3/.00621,-2.5/.02275,-2
.06681,-1.5/.11507,-1.2/.15866,-1/.21186,-.8/.27425,-.6
.34458,-.4/.42074,-.2/.5,0/.57926,.2/.65542,.4
.72575,.6/.78814,.8/.84134,1/.88493,1.2/.93319,1.5
.97725,2/.99379,2.5/.99865,3/.99997,4/1.5
 DMAND  FUNCTION   RN1,C11
0,0/.013,1000/.051,2000/.115,3000/.207,4000/.327,5000/.471,6000
.634,7000/.804,8000/.946,9000/1,10001
 ALT3   TABLE      X54,-300,50,200
 AVAR   FVARIABLE  (X10*V$NPV-K300000)/K1000
 BVAR   FVARIABLE  (X10*V$NPV-K750000)/K1000
 AMAX   FVARIABLE  (K3000*V$NPV-K300000)/K1000
 NPV    FVARIABLE  (F$PRESV*K30)+K200
        RMULT      ,71
        GENERATE   ,,,1
 LOOP1  SAVEVALUE  10,FN$DMAND
        TEST G     X10,K100,LOOP2
        SAVEVALUE  54,V$BVAR
        TABLULATE  ALT3
        TRANSFER   .LOOP3
 LOOP2  TEST LE    X10-K3000,LOOP4
        SAVEVALUE  54,V$AVAR
        TABLUATE   ALT3
        TRANSFER   ,LOOP3
 LOOP4  SAVEVALUE  54,V$AMAX
        TABLUATE   ALT3
 LOOP3  SPLIT      1,LOOP1
        TERMINATE  1
        START      1000
        CLEAR
        INITIAL    X100,4400
        START      1000
        CLEAR
        INITIAL    X100,4600
        START      1000
        CLEAR
        INITIAL    X100,4800
        START      1000
        CLEAR
        INITIAL    X100,5200
        END
```

Figure 3-12. GPSS simulation program for alternative 3 (stochastic event) of modification of Section 3.5.

the criterion used has been cost. One should not underrate the single criterion approach, however, because it is a useful step on the road to the harder multiattribute problem. By way of definition, multiattribute decision problems deal with choosing among a set of alternatives that are described in terms of their attributes. (Other terms sometimes used instead of "attribute" are: characteristic, aspect, factor, component, etc.) For example, Table 3-3 lists three alternatives, i.e., three different types of resource recovery systems under consideration for a particular municipality. Admittedly oversimplified,

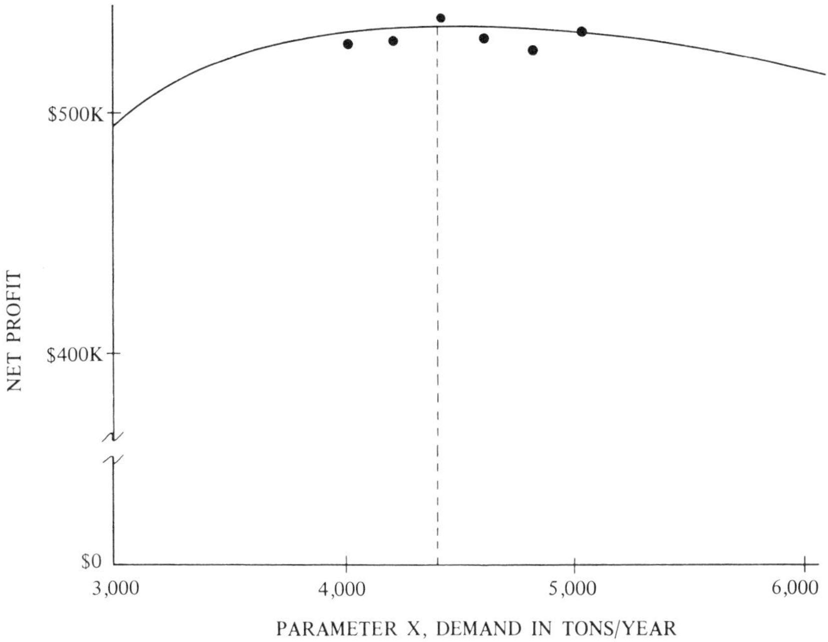

Figure 3-13. Results of simulations of "build small plant, enlarge if demand exceeds X ton/yr."

Table 3-3. Summary of Attributes for Decision Problem of Section 3.6

	Alternatives		
Attribute	**A_1**[a]	**A_2**[b]	**A_3**[c]
1. Process Cost/Ton	$15	$20	$30
2. System Reliability	0.7	0.8	0.9
3. Nuisance Potential	Some air pollution in the form of dusts.	Some water pollution in the form of dissolved solids and other materials; Cl and Pb air emissions.	Closed system—no pollution but some odors.
4. Market Reliability	Many potential industrial customers can be developed for fluff fuel.	Single-steam customer.	Alcohol can be sold on existing market to many customers.

[a]A_1 = dry MSW processing plant producing shredded RDF.
[b]A_2 = mass-burning MSW plant producing steam.
[c]A_3 = wet MSW processing plant producing ethanol.

four attributes have been identified as criteria for decision-making: process cost/ton, system reliability, nuisance potential and market reliability. The reader should understand that the attributes are assumed to be independent, e.g., process cost/ton includes the effects of system reliability for any given system; however, system reliability has effects other than just process cost. A system with much downtime may induce certain health hazards that are not computed as part of the traditional cost/ton figure. The multiattribute decision problem involves (1) the decision-maker's preference among values of a given attribute (how much does he prefer a $15/ton cost over a $20/ton cost), and (2) the decision-maker's preference across attributes (how much more important is process cost/ton over nuisance potential).

Presuming that measurements can be made on each attribute, then the problem of comparing one alternative with another is identical to that of comparing one vector with another. In the case of our three alternatives problem, the vectors might look like the following:

$$A_1\text{: } (15, 0.7, M, M)$$
$$A_2\text{: } (20, 0.8, H, H)$$
$$A_3\text{: } (30, 0.9, L, L)$$

where the A_i refers to the ith alternative, and L, M and H refer to (in some quantitative sense) low, medium and high desirabilities of the value of the attribute indicated. (The position of the elements of the vector follows the order of the attributes given in Table 3-3.) An alternative A_i is said to "dominate" alternative A_j if A_i is either equal or superior to A_j for every attribute, *and* for at least one attribute is superior to A_j. If one vector dominates the other, there is no problem in choosing between the alternatives because the basis for selection need be only a single attribute (i.e., any attribute for which the measurements on the two waste streams differ). As an example, suppose there was a fourth alternative as follows:

$$A_4\text{: } (30, 0.9, M, L)$$

Clearly, alternative A_4 is superior to A_3, and the decision to select A_4 over A_3 could be made simply on the basis of the third attribute–nuisance potential. However, among the alternatives A_1, A_2 and A_3, there is no dominance exhibited. The solution followed in this section is to convert each vector of measurements into an overall "score," "rating" or "index" on some interval scale. On an interval scale, not only are the positions arranged in terms of greater, equal or less of a given attribute, the units (or intervals) of measurement are equal. Thus, interval scales permit inferences to be made about the differences between the alternatives.

The procedure is as follows:

1. Rate or score each alternative against the criterion, assigning a numerical value to each alternative–criterion pair.

2. Combine these ratings within an alternative according to some mathematical rule.
3. Use the resulting combined scores of the alternative to select the alternative used.

Following this procedure, suppose we have assigned the numerical ratings shown in Table 3-4 for our problem, the scale being such that the higher the number, the more attractive the alternative. According to this procedure at least, the order of decreasing value of the alternatives is A_2, A_1 and A_3. Among the many questions that are provoked by this procedure we shall consider the following three:

1. *Question 1* (Unidimensionality). What justification is there for using a single number for each alternative total in Table 3-4 rather than a vector of numbers?
2. *Question 2* (Additivity). What justification is there for using an additive procedure for obtaining the total value of each alternative as shown in Table 3-4?
3. *Question 3* (Assessment). How does one go about assigning the rating values shown in Table 3-4?

Unidimensionality Question

It would be necessary to review several elements of utility theory [4] to provide a full explanation for the answer to the first question. Consequently, only the results will be given here. The relevant portion of the theory [5] states roughly that we assume that no criterion is infinitely more desirable or undesirable than some other criterion. We may prefer a process cost of $10/ton to a nuisance potential of "medium," for example, but we do not prefer it infinitely more. This property is known as "unidimensionality." If this property did not hold (i.e., multidimensionality), then alternatives would be evaluated solely on the most preferred criterion. Only if all alternatives were equal in this attribute would the next most preferred criterion enter into the consideration of alternatives. In Table 3-3, for example, if the most preferred criterion (i.e., the criterion infinitely most preferred to all others) was cost, the ranking (in order of preference) of the first two alternatives

Table 3-4. Rating Values, Example of Section 3.6

Alternative	Attributes				Totals
	Process Cost/Ton	System Reliability	Nuisance Potential	Market Reliability	
A_1	46	0	6	11	63
A_2	41	25	0	0	66
A_3	0	31	8	15	54

would be A_1-A_2. Suppose, however, that the cost value for alternative A_1 were \$20, the same as for A_2. If system reliability was most preferred (infinitely) to the nuisance potential and market reliability criteria, the ranking of the first two alternatives would be A_2-A_1, the rating of the two alternatives being based on system reliability because of the tie on the cost criterion. Although there are instances in which one would argue that there is no change in criterion x that could compensate for a change in criterion y, such instances do not appear to be common.

Finally, it should be noted that multiattribute utility and multidimensional utility are two different concepts. The former refers to the number of criteria involved in a decision, the latter to how the decision-maker relatively values those criteria.

Additivity Question

Utility theory also provides the answer to the second question. If the criteria are independent, then the additive procedure of Table 3-4 is valid; if they are interdependent, the additive procedure has no meaning [6]. By "independence" is meant that the utility rating of a basic criterion measurement within a waste stream does not depend on the others. As an example in which independence fails, consider the selction of a "snack" consisting of two attributes, a beverage and a food. Suppose there are two beverages, beer and milk, and two foods, pretzels and cookies. The choice of beverage here is hardly likely to be independent of the choice of food.

Assessment Question

The remainder of this section primarily will involve the assessment question. First we note that Table 3-4 can be put into the "weighted" form shown in Table 3-5. The numbers in a column of Table 3-4 can be obtained by multiplying the numbers in the corresponding column of Table 3-5 by the

Table 3-5. Transformed Version of Table 3-4

	Attributes				
Alternative	Process Cost/Ton 0.46	System Reliability 0.31	Nuisance Potential 0.08	Market Reliability 0.15	Totals
A_1	100	0	80	70	63
A_2	90	80	0	0	66
A_3	0	100	100	100	54

weighting factor at the head of that column. To get the total for rows in Table 3-5, the row elements are first multiplied by their respective weighting factors and then summed. (In point of fact, the weighing factors are nothing more than the relative utilities for the criteria.)

Although Table 3-4 must be derived by the individual to rate any given alternative, it is a complex and confusing task to derive a utility function simultaneously for more than a few criteria. A simpler approach is to work with Table 3-5. Here we are confronted with two tasks: The first is to evaluate the elements of a given criterion (X_i) relative to the other elements of X_i. Assuming independence, this may be done separately for each X_i. For a given X_i we may (arbitrarily) assign the number 100 to the most desirable element of X_i and assign the number 0 to the least desirable element. The intermediate elements of X_i are assigned a number between 0 and 100 according to the individual's estimate of their value relative to the least and most desirable elements. The results of this first task produce the numbers shown in the body of Table 3-5.

The assignment of such intermediate elements for a given criterion is nothing more than the derivation of a single attribute utility function. One well-known methodology for quantifying these utilities [7] utilizes a lottery approach. For a given criterion, consider the following lottery:

$$L_1 = \left\{ \begin{array}{c} p \text{ chance of least desirable element} \\ 1 - p \text{ chance of most desirable element} \end{array} \right\} . \tag{3-2}$$

"Lottery" here means that the decision-maker, if he accepts the lottery, will have a p chance ($0 < p < 1$) of obtaining the least desirable element and a $1 - p$ chance of obtaining the most desirable element. (If $p = 0.3$, e.g., we place into a jar three red balls and seven black balls and mix them thoroughly. A ball would be selected at random. If it was red, the decision-maker would get the least desirable element; if black, he would get the most desirable element.) We now consider another lottery:

$$L_2 = \{\text{intermediate element in question for certain}\} \tag{3-3}$$

and ask the decision-maker at what value of p he would be indifferent between the two lotteries. If he is indifferent between the two lotteries, then their expected utilities must be equal, i.e.,

$$u(\text{intermediate}) = (p)u(\text{most}) + (1 - p)u(\text{least}) \ . \tag{3-4}$$

Since we have arbitrarily assigned values to u(most) and u(least), u(intermediate) is readily determined, e.g., for cost/ton the most and least desirable elements are \$15/ton and \$30/ton, respectively. We have arbitrarily assigned the utility values of 100 and 0 to these elements. Suppose now the decision-maker says he is indifferent between

$$L_1 = \begin{Bmatrix} 0.9 \text{ chance that cost/ton} = \$15 \\ 0.1 \text{ chance that cost/ton} = \$30 \end{Bmatrix}$$

and

$$L_2 = \{\text{cost/ton} = \$20 \text{ for certain}\}$$

Then

$$\begin{aligned} u(\$20/\text{ton}) &= 0.9u(\$15/\text{ton}) + 0.1u(\$30/\text{ton}) \\ &= 0.9(100) + (0) = 90 \end{aligned}$$

Proceeding, we establish the numbers in the body of Table 3-5.

The second task is to establish the weighting factors using a method derived by the author called "DARE" [8]. First, the criteria are listed in arbitrary order, as shown in Table 3-6. Secondly, the range of each criterion in the problem is listed. The ranges can be either quantitative (e.g., process cost/ton) or descriptive (e.g., nuisance potential). The numbers in the R-column are quantitative ratios, assigned by the decision-maker, reflecting a comparison of the relative importance of the *range* of each criterion with the range of the one listed immediately below (e.g., the system reliability range is considered to be four times as important as the nuisance potential range; the nuisance potential range is considered to be one-half as important as the market reliability range, etc.). Once these ratios are determined by the decision-maker, the values of the weighting factors are fixed on a ratio scale. The R-column forms part of a simple algorithm that determines the values of the weighting factors. The integer 1 is entered in the F_i column opposite the last criterion. The remaining F_i entries are formed by the product of its predecessor and the R-value opposite it in a sort of "zigzag"

Table 3-6. Determination of Weighting Factors

Criteria	Attribute Range	R_i	F_i	W_i
Process Cost/Ton	$15	1.5	3	0.46
System Reliability	0.2	4	2	0.31
Nuisance Potential	From no pollution (some odors) to water (dissolved solids) and air (Cl and Pb) pollution.	0.5	0.5	0.08
Market Reliability	From single-steam customer to many customers on existing market.	–	1	0.15
		Totals	6.5	1.00

route up the table (e.g., the next-to-last element in the F_i column, 0.5, was formed by multiplying the 1 below it by the 0.5 to its left; the next F_i column element, 2, was formed by multiplying the 0.5 below it by the 4 to its left, etc.). The mathematical justification for the DARE procedure is given by Klee [8].

Although the order of listing of the criteria in Table 3-6 is arbitrary, it is assumed that the decision-maker is quantitatively consistent in his judgments (e.g., if A is two times as important as B, and if B is three times as important as C, then to be consistent, A must be six times as important as C). This seldom occurs in practice, however.

An explicit measure of consistency has been developed by the author as follows. First, an additional R-value, R_n, is estimated, linking the last criterion to the first one. For the criteria of Table 3-6, for example, one would estimate the relative importance of the range of the market reliability criterion to the range of the process cost per ton. To be consistent in the input data of Table 3-6, this R-value would have to be one-third because

$$R_i = W_i/W_{i+1} \tag{3-6}$$

$$R_n = W_n/W_{n+1} = W_n/W_1 \tag{3-7}$$

$$\prod_{i=1}^{n} W_i/W_{i+1} = 1 \tag{3-8}$$

In other words, applying Equation 3-8 to the data of Table 3-6,

$$(W_1/W_2)(W_2/W_3)(W_3/W_4)(W_4/W_1) = (3/2)(4)(1/2)(1/3) = 1$$

We now define a measure of consistency, C, as follows:

$$C = 1 - \exp[\sqrt{n}\, P \ln(1 - P)] \tag{3-9}$$

where n is the number of criteria and $P = \prod W_i/W_{i+1}$, if $\prod W_i/W_{i+1} \leqslant 1$ and $1/(\prod W_i/W_{i+1})$ if $\prod W_i/W_{i+1} > 1$. Note that C has the following properties: (1) when P = 1, C = 1; (2) when P = 0, C = 0; and (3) C increases with n.

For example, a moderate inconsistency would be if $R_4 = 1/5$. Then,

$$\prod W_i/W_{i+1} = (3/2)(4)(1/2)(1/5) = 3/5$$

and

$$C = 1 - \exp[\sqrt{4}\,(3/5)\ln(2/5)] = 0.67$$

If we introduce a substantial inconsistency, say $R_4 = 2$, then

$$\prod (W_i/W_{i+1}) = (3/2)(4)(1/2)(2) = 6$$

and

$$C = 1 - \exp[\sqrt{4}\,(1/6)\ln(5/6)] = 0.06$$

If the decision-maker finds C to be low (say below 0.9), then the individual R_i must be reexamined and adjusted so that Equation 3-8 is satisfied.

The weighting factors can be estimated, even when inconsistency is present, following a procedure devised by Saaty [9]. First we define a matrix, **R**, consisting of the R-values as follows:

$$\mathbf{R} = \begin{bmatrix} W_1/W_1 & W_1/W_2 \ldots W_1/W_n \\ W_2/W_1 & W_2/W_2 \ldots W_2/W_n \\ \cdot & \\ \cdot & \\ \cdot & \\ W_n/W_1 & W_n/W_2 \ldots W_n/W_n \end{bmatrix} \quad (3\text{-}10)$$

Let us consider the product of R and the vector W consisting of the weights

$$\mathbf{W} = \begin{bmatrix} W_1 \\ W_2 \\ \cdot \\ \cdot \\ \cdot \\ W_n \end{bmatrix} \quad (3\text{-}11)$$

$$\mathbf{RW} = \begin{bmatrix} n\,W_1 \\ n\,W_2 \\ \cdot \\ \cdot \\ \cdot \\ n\,W_n \end{bmatrix} = n \begin{bmatrix} W_1 \\ W_2 \\ \cdot \\ \cdot \\ \cdot \\ W_n \end{bmatrix} = n\,\mathbf{W} \quad (3\text{-}12)$$

Therefore, n is an eigenvalue of R, i.e., n satisfies the equation

$$\det(\mathbf{R} - \lambda \mathbf{I}) = 0 \quad (3\text{-}13)$$

and **W** is an eigenvector associated with that eigenvalue. If the **R**-matrix is consistent, then it can be shown that **R** has n − 1 eigenvalues that are zero and the remaining eigenvalue, λ_{max}, is n. For example, assuming the values in Table 3-6 of $W_1/W_2 = 3/2$, $W_2/W_3 = 4$, $W_3/W_4 = 1/2$, and the consistent value, $W_4/W_1 = 1/3$, the eigenvalues of

$$\mathbf{R} = \begin{bmatrix} 1 & 3/2 & 6 & 3 \\ 2/3 & 1 & 4 & 2 \\ 1/6 & 1/4 & 1 & 1/2 \\ 1/3 & 1/4 & 2 & 1 \end{bmatrix}$$

are 4, 0, 0 and 0. Furthermore, the eigenvector associated with $\lambda = 4$ is

$$\mathbf{W} = \begin{bmatrix} 1.14 \\ 0.76 \\ 0.19 \\ 0.38 \end{bmatrix}$$

By normalizing **W** to the unit vector (i.e., dividing each element of **W** by the sum of its elements), we obtain

$$\mathbf{W_n} = \frac{1}{2.47} \begin{bmatrix} 1.14 \\ 0.76 \\ 0.19 \\ 0.38 \end{bmatrix} = \begin{bmatrix} 0.46 \\ 0.31 \\ 0.08 \\ 0.15 \end{bmatrix}$$

which are the weights obtained previously in Table 3-6.

Now let us suppose that the decision-maker was moderately inconsistent, i.e., $R_4 = 1/5$. The maximum eigenvector of the revised R-matrix,

$$\mathbf{R} = \begin{bmatrix} 1 & 3/2 & 6 & 5 \\ 2/3 & 1 & 4 & 2 \\ 1/6 & 1/4 & 1 & 1/2 \\ 2 & 1/2 & 2 & 1 \end{bmatrix}$$

is 4.03, and the associated unit eigenvector is

$$\mathbf{W_n} = \begin{bmatrix} 0.51 \\ 0.29 \\ 0.07 \\ 0.13 \end{bmatrix}$$

Let us introduce even more inconsistency by letting $R_4 = 2$. The maximum eigenvector of the once more revised **R**-matrix,

$$\mathbf{R} = \begin{bmatrix} 1 & 3/2 & 6 & 1/2 \\ 2/3 & 1 & 4 & 2 \\ 1/6 & 1/4 & 1 & 1/2 \\ 2 & 1/2 & 2 & 1 \end{bmatrix}$$

is 4.43 and the associated unit eigenvector is

$$\mathbf{W_n} = \begin{bmatrix} 0.32 \\ 0.32 \\ 0.08 \\ 0.28 \end{bmatrix}$$

As more and more inconsistency is introduced into our set of R-values, the greater the deviation of λ_{max} from n and of the estimated weighting factors from the true ones. It has been suggested that λ_{max} or functions of λ_{max} be used as a measure of consistency [10], but the author finds these methods rather insensitive to substantial changes in consistency, so recommends the measure of Equation 3-9.

Whether the eigenvalue method has any real advantage over the DARE method for the estimation of the weighting factors is doubtful. Eigenvalue and eigenvector calculations must be done using a digital computer, which further complicates matters and, when inconsistency is present, the author sees no value in calculating weighing factors that clearly are incorrect. The solution to the problem is to remove the inconsistency in the first place; if this is done, the DARE method will produce the correct weighting factors utilizing the simple hand calculations illustrated in Table 3-6.

Finally, we have shown how the three alternatives in our example problem have been rated. Alternative A_2, with a score of 66, is superior to alternatives A_1 (with a score of 63) and A_3 (with a score of 54). Note that these ratings are defined on an interval rather than a ratio scale. It is not true, for example, that since in Table 3-5 alternative A_2 is rated at 66 and A_3 at 54, alternative A_2 is 66/54 = 1.22 or 22% better than A_3. It is, rather, the relative differences that are preserved. For example, the difference (3) between A_1 and A_2 is one-fourth the difference (2) between A_2 and A_3. Thus, there is four times the improvement in going from A_3 to A_2 than from A_1 to A_2.

3.7 COMMENTS ON LINEAR DECISION MODELS

The reader will observe that the assumptions of the methodology described in Section 3.6 produce a linear model, i.e., the final ratings of Table 3-5 are of the form

$$S_j = \sum_i W_i u_{ij} \tag{3-14}$$

where S_j is the score for the jth alternative, W_i is the weight of the ith criterion, and u_{ij} is the utility value of the jth alternative on the ith criterion. Deviations from the assumptions may be supposed to result in a nonlinear model for the true state of affairs. The question is, then, how sensitive is the model to the assumptions of the method? The answer is that linear models are exceedingly "robust" when correlation with a criterion is the figure of merit [11]. For example, suppose that

$$f = xy \tag{3-15}$$

i.e., a multiplicative model. How close an approximation is the following linear function?

$$g = x + y \qquad (3\text{-}16)$$

To make the computations simpler, we assume that x and y have been normalized so that the region of interest is the unit square. Assume that x and y are uniformly distributed over the unit square. (This is the least favorable assumption because, as Dalkey [11] points out, any linear correlation between x and y would increase the correlation between f and g.) The correlation between f and g, however, is 0.926. (The proof is given in Appendix A.1.) As another example, suppose that

$$f = ax + by \qquad (3\text{-}17)$$

and

$$g = cx + dy \qquad (3\text{-}10)$$

with a = 1, b = 2, c = 2 and d = 1. The correlation between f and g in this case is 0.8. (This proof is also given in Appendix A.2.) Hence, the error-free correlations are sufficiently high so that it would be questionable whether the noisy kind of data (i.e., uncertainty of input) met in practice could distinguish the models within each of these examples.

3.8 SUMMARY

Uncertainty analysis takes into account precisely what its name suggests, i.e., the fact that there is little or no assurance that our inputs to a cost or profit calculation that affect a decision are error-free. Uncertainty analysis, coupled with simulation, allows us to take into consideration particularized uncertainty distributions for the various factors. Using simulation, we are not forced to accept "classical" distributions merely because they are mathematically tractable. Furthermore, it automatically takes into account complicated interactions between factors and provides us with information that conventional techniques cannot provide, such as the probability of losing money, the degree of uncertainty of the final output, etc.

When the problem is one of sequential decision-making (i.e., the analysis of a number of highly interrelated decisions occurring at different points in time), a slightly different approach must be taken. The approach recommended here is the decision tree, modified to accept any stochastic event that serves to reflect the real world. The single-valued traditional engineering economic analysis is indeed outdated and dysfunctional. The "firm" numbers produced by this kind of analysis are misleading and guaranteed to be incorrect. Decisions should be made knowing as much about the risks involved as possible. If information is available, it should be used, not swept under the rug of either convenience or tradition.

Finally, with regard to the multiattribute decision model, in addition to providing a workable methodology, the technique makes clear (1) what criteria are used to evaluate the alternatives; (2) what is the evaluator's view of the relative importance of the criteria; and (3) what is the worth of each alternative with respect to each criterion. It has the virtue of all explicit methods in that it is subject to review by others to see whether any mistakes have been made in the application of the method–in calculations or manipulation of the data.

APPENDIX TO CHAPTER 3

A 3.1 CORRELATION OF MULTIPLICATIVE WITH LINEAR MODEL

We assume that x and y are uniformly but independently distributed over the unit square. Thus, the expectation of x and y, E(x) and E(y), are each equal to 1/2, and their standard deviations are each equal to 2/7 [12]. Since $E(A^2) = var(A) + [E(A)]^2$, $E(x^2) = E(y^2) = 65/196$. Let f = xy and g = x + y. The correlation between any two functions, f and g, is [13]:

$$\text{rho}(f,g) = [E(fg) - E(f)E(g)]/\sqrt{\{E(f^2) - [E(f)]^2\}\ \{E(g^2) - [E(g)]^2\}} \tag{A-1}$$

Because of the independence between x and y,

$$E(f) = E(xy) = E(x)E(y) = 1/4 \tag{A-2}$$

$$E(g) = E(x + y) = E(x) + E(y) = 1 \tag{A-3}$$

$$E(f^2) = E(x^2y^2) = E(x^2)E(y^2) = (65/196)^2 \tag{A-4}$$

$$E(g^2) = E(x^2 + 2xy + y^2) = E(x^2) + 2E(x)E(y) + E(y^2) = 228/196 \tag{A-5}$$

$$E(fg) = E[(xy)(x + y)] = E(x^2y + xy^2) = E(x^2)E(y) + E(x)E(y^2) = 65/196 \tag{A-6}$$

Substituting the values of A-2 through A-6 into A-1, we have

$$\text{rho}(f,g) = [(65/196) - 1/4]/\sqrt{[(65/196)^2 - 1/16]\,[(228)/196) - 1]} = 0.926$$

A 3.2 CORRELATION OF TWO LINEAR MODELS

Suppose now that f = ax + by and g = cx + dy, and that we assume the same restrictions on x and y as in Section A 3.1. Equation A-1 can be put into the following form:

$$\text{rho}(f,g) = \text{cov}(f,g)/\sqrt{\text{var}(f)\text{var}(g)} \tag{A-7}$$

Now, from the Propagation of Error formula (Equation 1-9, Chapter I),

$$\text{cov}(f,g) = (ac + bd)\sigma^2 \qquad \text{(A-8)}$$

$$\text{var}(f) = (a^2 + b^2)\sigma^2 \qquad \text{(A-9)}$$

$$\text{var}(g) = (c^2 + d^2)\sigma^2 \qquad \text{(A-10)}$$

where σ^2 = var(x) = var(y). Substituting the values of A-8 through A-10 into Equation A-7, we have

$$\text{rho}(f,g) = (ac + bd)/\sqrt{(a^2 + b^2)(c^2 + d^2)} \qquad \text{(A-11)}$$

When a = 1, b = 2, c = 2 and d = 1, then

$$\text{rho}(f,g) = (2 + 2)/\sqrt{(1 + 4)(4 + 1)} = 0.8$$

REFERENCES

1. Brown, R. V., A. S. Kahr and C. Peterson. *Decision Analysis For The Manager* (New York: Holt, Rinehart and Winston, 1974).
2. Magee, J. F. "Decision Trees for Decision Making," *Harvard Bus. Rev.* 42:135 (July-August 1964).
3. Hespos, R. F., and P. A. Strassmann. "Stochastic Decision Trees For The Analysis of Investment Decisions," *Management Sci.* 11(40):B244-B259 (August 1965).
4. Von Neumann, J., and O. Morgenstern. *Theory of Games and Economic Behavior*, 2nd ed. (Princeton, NJ: Princeton University Press, 1947).
5. Fishburn, P. C. *Modern Utility Theory: 1940-1970*, Defense Documentation Center, AD 625047 (1965), pp. 1-4.
6. Fishburn, P. C. "Independence in Utility Theory with Whole Product Sets," *Operations Res.* 13(1):28-45 (1965).
7. Bierman, H. et al. *Quantitative Analysis for Business Decisions* (Homewood, IL. Richard D. Irwin, Inc., 1965), pp. 198 201.
8. Klee, A. J. "The Role of Decision Models in the Evaluation of Competing Environmental Health Alternatives," *Management Sci.* 18(2):52-67 (1971).
9. Saaty, T. L. "A Scaling Method for Priorities in Hierarchical Structures," *J. Math. Psychol.* 15:234-281 (1977).
10. Yager, R. R. "An Eigenvalue Method of Obtaining Subjected Probabilities," *Behavioral Sci.* 24:382-387 (1979).
11. Dalkey, N. C. "A Delphi Study of Factors Affecting the Quality of Life," in *The Delphi Method: Techniques and Applications*, H. A. Linstone and M. Turoff, Eds. (Reading, MA: Addison-Wesley Publishing Co., Inc., 1975), pp. 387-401.
12. Natrella, M. G. "Characterizing the Measured Performance of a Material, Product, or Process," in *Experimental Statistics*, National Bureau of Standards Handbook 91, Washington, DC (1963), pp. 2-9.
13. Hogg, R. V., and A. T. Craig. "Multivariate Distribution," in *Introduction to Mathematical Statistics* (New York: Macmillan Publishing Co., Inc., 1959), pp. 211,212.

CHAPTER 4

THE UTILIZATION OF EXPERT JUDGMENT

4.1 INTRODUCTION

In many decision-making situations, the use of experts plays a vital role. The reasons are not hard to find. The decision-maker, for example, may himself have very little knowledge of the pertinent subject matter; alternatively, he may lack the analytical capacity to evaluate the problems before him. In the inexact sciences (and this certainly includes the environmental sciences), the vagueness of certain background knowledge is typical (even though it may be significant or even predominantly important) as is the uncertainty about the evidence accorded the various kinds of available information.

As Helmer and Rescher [1] suggest, "Hence the great importance which must be attached to experts and to expertise in these fields. For the expert has at his ready disposal a large store of (mostly inarticulated) background knowledge and a refined sensitivity to its relevance, through the intuitive application of which he is often able to produce trustworthy personal probabilities, regarding hypotheses in his area of expertness."

Thus, there is a basic assumption that underlies the use of expert judgment. In situations of uncertainty (incomplete information or inadequate theories), expert judgment can be used as a surrogate for direct knowledge. This general assumption is hardly in doubt. What is in doubt is the degree of accuracy and other qualities of such estimates.

We may hypothesize that a set of events exists in which the events may be either discrete or continuous, and that we require from our expert his estimate for each event. In many cases, the estimate is merely a selection from the set. For example, the expert might say, "It is not economically feasible to recover glass," a selection from the two-event set "economically feasible" and "not economically feasible." Or our expert might forecast, "National beverage container legislation will be passed by 1985," a selection of a single date out of a continuum. Thus, the expert implicitly chooses a probability of either 0 or 1 for each event.

On the other hand, the expert might be required to make a probability estimate for any or all of these events. These estimates can be probability assignments for discrete alternatives (e.g., a probability of 0.7 that legislation is passed in the interval 1985-1990) or continuous distributions for continuous quantities (e.g., a probability distribution of solid waste quantity generated per capita by 1990). Another related kind of estimate is the estimation of the functional relationship between two or more variables, such as the slope of a linear trend; these may be either point estimates or discrete or continuous probability distributions.

What constitutes an expert is an important question, and several criteria have appeared in the literature. Dalkey [2] has summarized a number of such criteria, and the following are of sufficient interest to mention here: definiteness, certainty, reliability, realism and honesty.

4.2 CRITERIA FOR EVALUATING EXPERTS

Definiteness

Definiteness refers to the degree of "sharpness" of the estimate. For a discrete probability distribution it refers to the degree to which the probabilities approach 0 or 1. There is an inherent attractiveness to an estimate in the case of discrete events that ascribes a probability of close to 1 for some alternative, and probabilities close to 0 to the others. There is a general feeling that probability estimates close to 0 or 1 are harder to make than "wishy washy" estimates in the neighborhood of 1/2. For a discrete probability distribution, definiteness can be measured by Σr_j^2, where r_j is the assessed probability for event j; for a continuous probability distribution, definiteness can be measured by the variance of the assessed distribution. Suppose, for example, in a discrete distribution involving these events, that the expert assigned a probability of 1/3 to each of them. Thus, $\Sigma r_j^2 = 1/3$. If he had assigned a probability of zero to the first two and one to the third, Σr_j^2 would equal 1. Figure 4-1 contrasts two continuous probability distribution estimates. The taller, narrower distribution is the more definite. In both cases the expert is predicting that per capita waste generation in 1990 will fall between 3 and 4 lb/person/day, with an expected value (average) of 5; the taller, narrow distribution is more definite, however.

Certainty

A related concept is certainty. In economics, a distinction is made between risk (i.e., situations involving probabilities but where the probabilities are

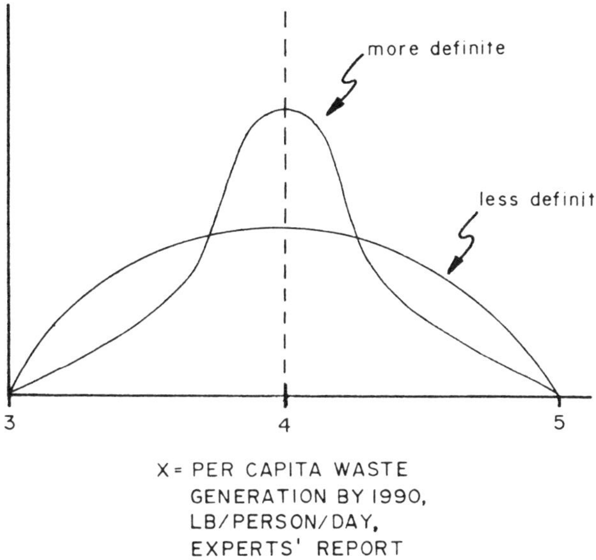

Figure 4-1. Definiteness.

known) and uncertainty (situations where the probabilities are not known). Consider, for example, the situation of two coins, where an individual is asked to estimate the probability of heads. Coin A is a common kind of coin where the individual has flipped it several times. In this case, he might say that the probability of heads is 1/2 with a high degree of confidence. Coin B, let's say, is an exotic object with an unconventional shape, and the individual has not flipped it at all. In the case of coin B he might also estimate a probability of 1/2 for heads, but he would be highly uncertain whether that is the actual probability. Therefore, probability 1/2 cannot express the uncertainty attached to the estimate for the second coin. To return to the example of the previous section, suppose our expert is required to produce a *point estimate*, Y, for per capita generation; we would expect this to coincide with the modal values from the distribution shown in Figure 4-2, i.e., 4 lb/capita/day. His belief in this value might be high or low and can be represented by this distribution in Figure 4-2. Note that the definiteness figure refers to his *reported* estimate, while the certainty figures refer to his beliefs. In other words, the distributions shown in Figure 4-1 represent risk, while the distributions in Figure 4-2 represent uncertainty.

That there is a relationship between definiteness and uncertainty is shown in Figure 4-3. As Y (the expert's belief) approaches an extreme value, the standard deviation of the distribution approaches 0. Assuming that the

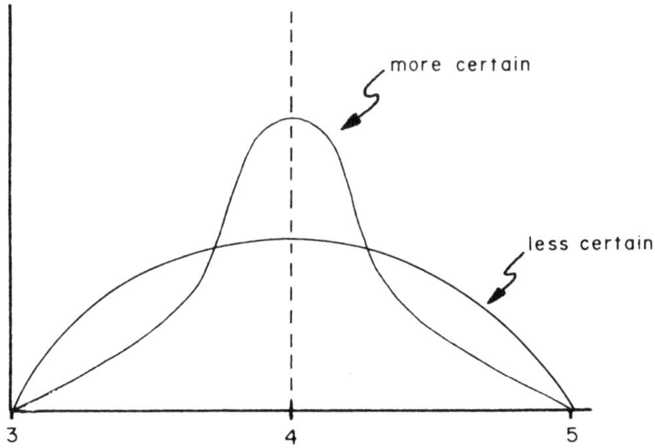

Figure 4-2. Certainty.

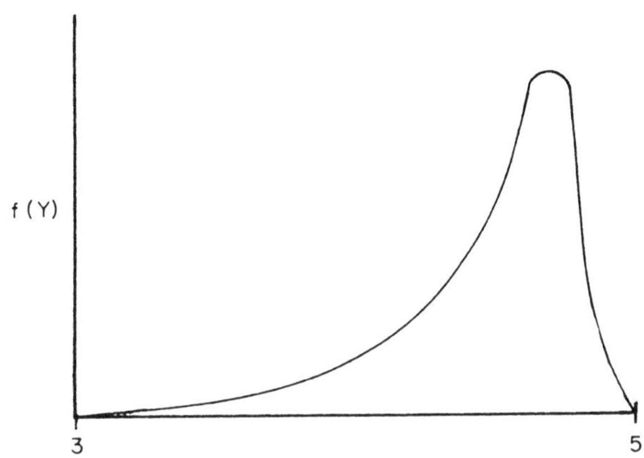

Figure 4-3. An extreme certainty distribution, which must have a small standard deviation.

expert gives the mean of his belief distribution as the mean of his report, then the standard deviation of his report distribution, X, also must approach 0. Under these circumstances, high certainty implies high definiteness. Unfortunately, if the mean of the belief distribution is around the midpoint of the range of Y, the standard deviation can be anything, as can be the standard deviation of the report distribution, X.

Finally, certainty can be ascertained by the use of self-ratings, which are discussed in Section 4.4.

Reliability and Realism

Reliability is the relative frequency of cases in which, when confronted with several alternatives, the expert ascribes to the actual event a greater probability than to the other events. The concept of reliability involves not just a single estimate, but rather a set of estimates generated by a single expert. For example, an expert might estimate that national beverage container legislation might be passed in 1980, 1981 or 1982, with probabilities of 0.2, 0.5 and 0.3, respectively. If such legislation is passed in 1981, this is a "success" and the proportion of successes among a set of such estimates constitutes the expert's reliability.

Consider now the specific estimate just described. Legislation did not occur in 1980; the expert had ascribed a probability of only 0.2 to this event so it was not likely to occur. Over the long run, however, in other estimates there most likely are some that involve 0.2 probabilities and some of these events will occur. If the concept of probability is being interpreted correctly, then 20% of the events to which a probability of 0.2 is ascribed should occur. Given enough data of this sort over a wide range of probabilities for events, a curve of the sort shown in Figure 4-4 could be generated. This is known as a "realism curve" and represents the relative frequency that the expert is correct, given the probability that the expert has estimated the event will occur. The straight line represents "perfect realism," and the curved line in Figure 4-4 is a typical curve in the sense that small probabilities are often underestimated and large probabilities often overestimated.

Honesty

It may not always be the case that the individual's reported estimate matches his actual belief. For example, group pressure may cause the individual to change his response to conform to the majority. The classic work of Asch [3] has generated much research dealing with this problem. In Asch's experiment, a naive subject was given the seemingly uncomplicated and

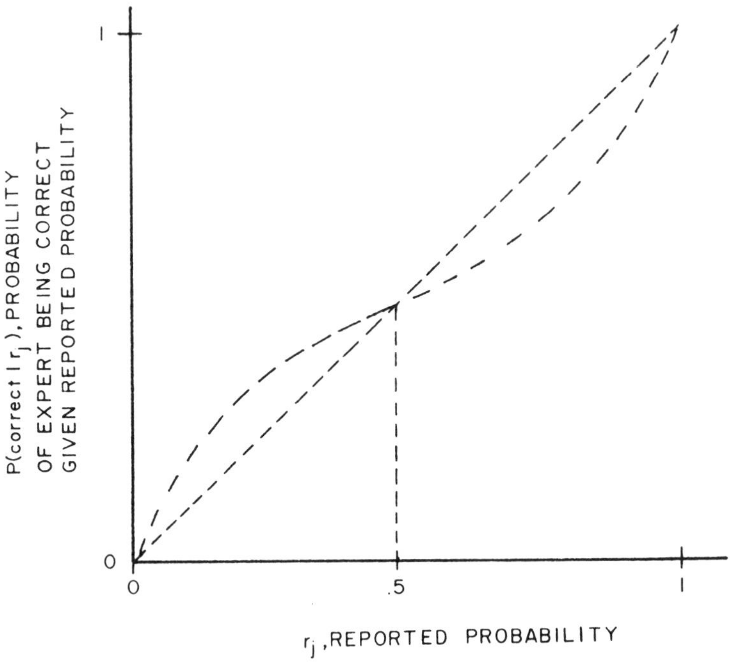

Figure 4-4. Example of realism curve (dotted line represents perfect realism).

straightforward task of making a series of judgments concerning which of three comparison lines is equal in length to a standard line. The subject answers after hearing the responses of several other people, ostensibly subjects like himself, but who deliberately give planned incorrect answers on certain critical trials. The subject's conformity score is the number of times he agrees and yields to the unanimous but incorrect majority. Among the results Asch reported were as follows:

1. When responding alone, all subjects achieve close to 100% accuracy.
2. When responding after the confederates, approximately one-third of the subjects yield and announce the incorrect majority response.
3. A unanimous majority of three elicits as many conforming responses as occur in groups as large as fifteen.
4. The effect of a single defection from the previously unanimous group is enough to induce significantly fewer conforming responses on the part of the naive subject.
5. There are very marked individual differences in susceptibility to group pressure, with some subjects consistently independent and others consistently yielding.

Methods for dealing with group pressure, such as Delphi, are described later in this chapter.

Another instance in which the expert's announced decision may not coincide with his true belief occurs when the expert is in the role of an advisor. In this case one may assume that he desires to attain goals of his own choosing. The expert, however, will not necessarily come closest to his own goal by stating it and arguing for it. If his advice is at all extreme from the decision-maker's point of view, he may find his recommendation written off as unrealistic. The expert's effectiveness and degree of goal achievement will then be low or even zero. On the other hand, if to sound realistic and maximize the probability of having his advice accepted the expert shades his goal to the point of making it the same as that of the decision-maker he serves, his degree of goal achievement, expressed as the difference between the decision-maker's views with and without the benefit of his advice, respectively, is also zero. In between, there is a maximum expected goal achievement. To illustrate this point, let us assume that the functional relationship between the probability of the expert's advice being accepted and the proportion by which he understates his true goal is linear. Further assume that the only possibilities are (1) acceptance of the expert's announced goal, or (2) total rejection, which will reduce its probability of acceptance to zero. This situation is shown graphically in Figure 4-5. The equation of this function is

$$P = 1 - F \tag{4-1}$$

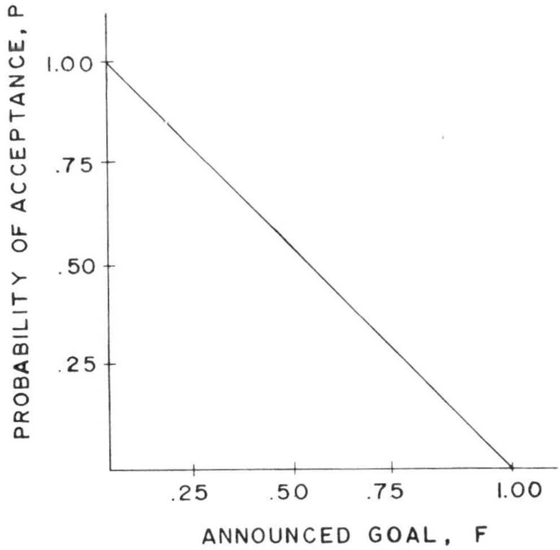

Figure 4-5. Probability, P, that the expert's advice will be accepted versus announced goal, F, expressed as a fraction of the difference between the decision-maker's views and the expert's true goal.

where P is the probability of acceptance and F is the expert's announced goal expressed as a fraction of the difference between the decision-maker's views and the expert's true goal.

The expected (average) goal achievement of the expert is his announced goal (that is, advice) F, times the probability that this advice will be accepted. In other words,

$$FP = F(1 - F) \tag{4-2}$$

By elementary calculus, we maximize FP by taking the first derivative of this function and setting it equal to zero, i.e.,

$$d(FP)/dF = 1 - 2F = 0 \tag{4-3}$$

Thus F = 1/2, i.e., the expert will maximize his goal achievement if he makes his announced goal equal to one-half the difference between his true goal and that of the decision-maker he serves. Different maxima will result if the expert, because of his greater or lesser powers of persuasion, can assume the relationship between goal announcement and probability of acceptance to be concave or convex, for example [4]. The implications for engineering consulting firms and for those who utilize such firms is that even the paid consultant might not always reveal his true judgments.

Another source of difference between true and stated belief is in how the expert is rated. Suppose we are scoring an expert on a probability assessment involving a discrete probability distribution where $p_1, p_2 \ldots, p_n$ are the true probabilities for the n values of the random variable, and $r_1, r_2 \ldots, r_n$ are the expert's assessed probabilities. Let the subscript h refer to the interval in which the random variable is actually observed. An intuitively reasonable scoring rule is to let $S = r_h$, i.e., the score is equal to the probability assigned to the event that actually occurs. It turns out, however, that the expert can maximize his expected score by setting one of his r_i equal to one, that is, the r_i equal to the highest p_i, and the other r_i equal to zero. Suppose $p_1 = 0.6$ and $p_2 = 0.4$ in an example where the random variable is "National Beverage Container Legislation Passed by 1985." The variable has two values, 0 and 1 (i.e., "no" and "yes"). What we are saying is that there is a 60% probability that it will not be passed by 1988 and a 40% probability that it will. By using the scoring rule $S = r_h$, the greater the probability the expert assigns to the event that actually occurs, the greater will be his score, i.e., he is rewarded for his definiteness. By setting $r_1 = 1$ and $r_2 = 0$, the expert's expected score is 1.0 (0.6) + 0.0 (0.4) = 0.60, whereas if he had stated what he actually believed (and assuming the expert is perfectly accurate), he would have stated $r_1 = 0.6$ and $r_2 = 0.4$, producing a score of 0.6 (0.6) + 0.4 (0.4) = 0.52. What is needed is a scoring rule that will *oblige* the expert to reveal his true judgments. Three such rules can be demonstrated [5]:

1. Quadratic scoring rule:

$$Q = 2r_h - \sum_{j=1}^{n} r_j^2 \tag{4-4}$$

2. Spherical gain scoring rule:

$$S = r_h \left[\sum_{j=1}^{n} r_j^2 \right]^{-1/2} \tag{4-5}$$

3. Logarithmic loss scoring rule:

$$L = \ln r_h \tag{4-6}$$

For example, our expert would have scored as follows under the three rules (where the "true report" probabilities are 0.6/0.4 and the "false belief" probabilities are 1/0, where the first number represents the assessed probability of the first event, and the second number represents the assessed probability of the second event):

1. Quadratic rule:

Expected Value True Report = 0.52

Expected Value False Report = 0.20

2. Spherical rule:

Expected Value True Report = 0.72

Expected Value False Report = 0.60

3. Logarithmic rule:

Expected Value True Report = −0.67

Expected Value False Report = −1.85

In all cases the score is higher for the true than for the false report. As an example of the calculations, the expected value of the true report for the quadratic rule is as follows:

$$0.6[2(0.6) - 0.36 - 0.16] + 0.4[2(0.4) - 0.36 - 0.16] = 0.52$$

Note: Since *ln* 0 is indefinite, r was taken as 0.99 and r_1 as 0.01 for the logarithmic rule calculation. Of course probabilities of 0 or 1 are not encountered in practice.

The realities of trying to implement any of these concepts are brought to us sharply by Savage [6]:

> No rule of this sort can claim absolute or objective validity. For example, actual past experience with the experts may be extensive, moderate, meager, or absolutely lacking. When past experience is extensive . . . the rule often has

> much to recommend it; when direct past experience with the experts is meager, the rule is silly; and when such experience is altogether lacking, the rule results in a tie and is therefore empty. Actually, if you have little or no past experience with the experts, you will have to ponder them in terms of whatever information it was that brought you to regard them as promising in the first place: this well finder is regarded by the whole neighborhood as infallible with the hazel fork; that one is a professor of geology and the author of an important treatise on subsurface hydrology but has never before tried to help anyone locate a well. In such a context, the subjective aspect of your decision is thrown into prominence, but no matter how much direct past experience you may have with the experts, the ultimate subjectivity of your choice among them never disappears, though its effects may become less agonizing. . . .

Finally, it ought to be noted that experts have often been wrong. Writing in 1889, Thomas A. Edison said: "There is no plea which will justify the use of high-tension and alternating currents, either in a scientific or a commercial sense." In a speech in the U.S. Senate, Senator Morris Sheppard, author of the 18th Amendment (Prohibition) commented: "There is as much chance of repealing the 18th Amendment as there is for a hummingbird to fly to the planet Mars with the Washington Monument tied to its tail." Hegel published his proof that there could be no more than seven planets just a week before the discovery of the eighth. In 1964 a panel of the National Academy of Sciences on Weather and Climate Modification called rainmaking by cloud-seeding astrology. In 1966 the same panel had to admit that there was something to rainmaking after all. It took 20 years from the invention of cloudseeding in 1946 for a consensus to be reached among scientists that "it works." In the interim, most of the advances that took place in the field occurred in spite of expert opinion.

4.3 ELICITING INDIVIDUAL JUDGMENT

There are times when there is no difficulty in drawing from the expert his judgment. He may have a very firm view of whether it is economically feasible to recover glass. On the other hand, he may be uncertain about the economics of glass recovery: he may have all the information needed to make a particular assessment but hesitates or appears uncertain because he has never thought about the question exactly as it has been posed. He may have certain prior knowledge that is not easy to express quantitatively. An elicitation technique may be helpful in drawing out the assessment from this prior knowledge.

One useful technique is based on the concept of "dichotomous search" [7]. Suppose we assume that national beverage container legislation will be passed sometime in the future and the problem is to obtain the expert's judgment of the year in which this will take place. The first step is to determine the earliest and latest date the event can take place. The exact boundaries are not too important; the only requirement is that the expert be in agreement. Suppose these extremes are 1980 and 1990. The next step is to divide the interval in half–1985 in this instance–and ask whether, in his judgment, the event will take place during the first interval (1980 to 1985) or the second (1988-1990). If he insists it will occur right on the boundary year, then of course we are finished. If not, we further divide the interval he has selected. If, for example, he selects 1980-1985, our mid-point is mid-1982, and again we ask him in which interval he thinks the event will occur. After n iterations, the interval of uncertainty (i.e., the last interval) will be of width $1/2^n$ of the original interval. If we start with an interval of eleven years, then in five iterations we will have the expert's estimate of the date to within approximately four months.

The same technique can be used to estimate a single probability. It can also be used to estimate a probability distribution. The first version to be described will be the cumulative probability (CP) method, which involves assessing fractiles (by means of either successive subdivisions or direct questions regarding fractiles) and graphing the cumulative probability distribution [8]. Let us illustrate the CP method by eliciting from our expert his judgment of the proportion, p, of households in a given city that would participate under stated circumstances in a source separation program for municipal solid waste. This is a number between 0 and 1. The following are the steps using the CP method and are represented by questions which we assume are answered by the dichotomous search technique:

1. Can you give me an estimate of p such that the true value is equally likely to fall below or above it? [Note that this is the 0.50 fractile, i.e., where the cumulative probability, F(p), equals 0.50.] (Suppose answer is 0.34.)
2. Now suppose you were told that p is less than 0.34. Can you determine a new point such that it is equally likely that p is less than or greater than this point? [Note that this is F(p) = 0.25.] (Suppose answer is 0.22.)
3. Now suppose you were told that p is less than 0.22. Can you determine a new point such that it is equally likely that p is less than or equal to this point? [Note that this is F(p) = 0.125.] (Suppose answer is 0.16.)

Continue this line of questioning to at least four subdivisions on each side of the median and graph the results as shown in Figure 4-6. Addition of intervals can be used to fill in "thin" areas of the curve. The resultant curve is the cumulative probability distribution for the expert's judgment of p, the proportion of householders willing to participate in source separation. If any of the data points appeared to be off a smooth curve drawn through the data, the expert would be asked to alter his judgments slightly to correct the situation. The expert is encouraged to examine the cumulative probability distribution and revise points until he is satisfied that it represents his true judgment and that existing inconsistencies are removed.

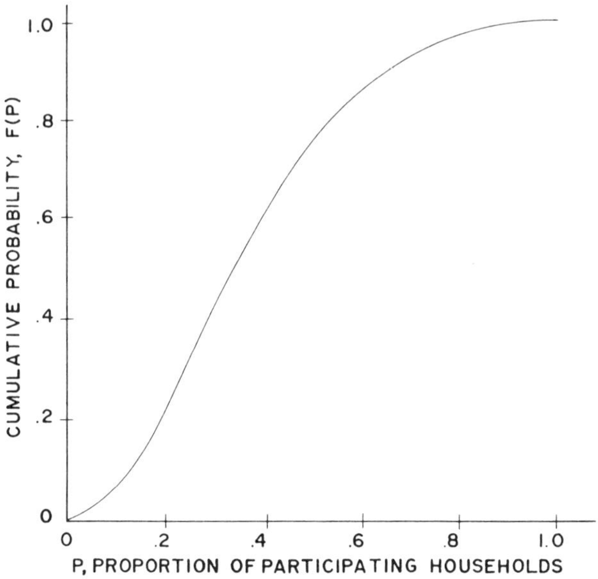

Figure 4-6. Elicitation of a probability distribution (CP method).

One inconsistency, for example, would be the construction of a cumulative probability distribution which implies a bimodal frequency distribution when it is the expert's real view that the frequency distribution is unimodal. Figure 4-7 shows two cases, i.e., an unimodal and a bimodal distribution, comparing the relative frequency distribution with its corresponding cumulative probability distribution. Note that the bimodal cumulative probability curve contains three points of inflection compared to the single point of inflection of the unimodal curve. If such curves are obtained, it should be ascertained that the expert really thinks that a bimodal (or multimodal) frequency distribution represents his assessment. If not, the inconsistency must be resolved and the cumulative probability distribution be redrawn.

One other approach is to have the expert establish the frequency distribution directly (i.e., the FD method). One starts off as described for the CP approach, but with the following modifications [9].

1. What do you consider to be the most likely value of p? Can you determine two values of p, one on either side of the most likely point, which are about half as likely as the most likely point? [Here it is explained that this value is the point where the frequency distribution assumes its maximum height (i.e., f(X) is at its maximum) and the half-as-likely values are points where the frequency distribution is half as high as it is at the point of maximum height. If necessary, a rough graph will illustrate the point.]

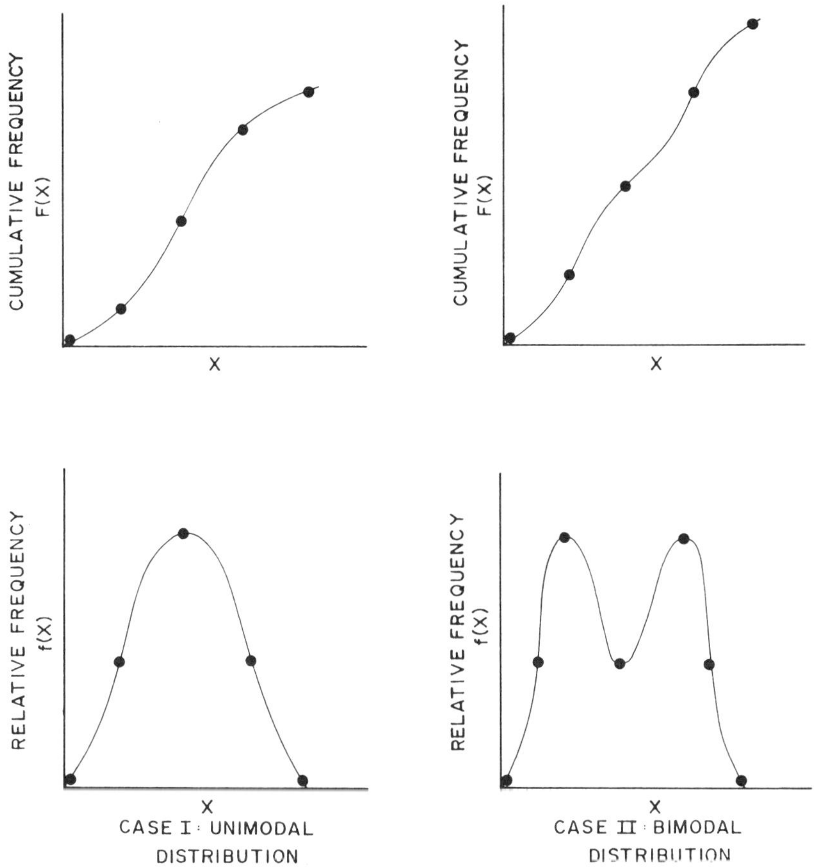

Figure 4-7. Relationship of relative frequency distribution to cumulative probability function.

2. Can you determine a value of p such that 1/2 the area under the graph of the frequency distribution is to the left of the point and 1/2 to the right?
3. Such that 1/4 of the area is to the left of the point and 3/4 to the right?
4. That 3/4 of the area is to the left of the point and 1/4 to the right?
5. Such that 1/100 of the area is to the left of the point and 99/100 to the right?
6. Such that 99/100 of the area is to the left of the point and 1/100 to the right?
7. Using the above-assessed point as your guide, can you draw on the graph what your frequency distribution looks like?

A frequency distribution, f(X), can be derived from a cumulative probability distribution, F(X), by constructing a histogram from the latter. Table 4-1 shows how this is done for the F(X) curve of Figure 4-6. Ten equal intervals of p from Figure 4-6 were selected and then corresponding F(p) values noted. Successive differences of F(p) produce the histogram bar heights, which are shown plotted in Figure 4-8. A smooth curve drawn through the histogram produces the required frequency distribution. It will be noted from Table 4-1 that the area of the histogram bars totals one, so that the frequency distribution is really a probability density function as well.

The reverse process can be used to produce a cumulative probability distribution from a frequency distribution, but a simpler method is frequently used by the author:

1. Draw vertical lines on the smooth curve of the frequency distribution at convenient intervals. This divides the distribution into vertical "slices." (Six to twelve slices will do.) Number the slices.
2. Cut out the frequency distribution with a pair of sharp scissors. Then cut the slices as well.
3. Weigh each slide on a sensitive balance and cumulate the weights.
4. Divide the cumulated weights by the total weight, and the result is the cumulative probability, which can be plotted against the x-value of the right-hand side of each slice. This is the cumulative probability distribution.

Since both the CP and FD are based on the same knowledge and judgments, and since the one can be transformed into the other equivalently, any differences between them amount to inconsistencies. Such discrepancies should be resolved by the expert himself.

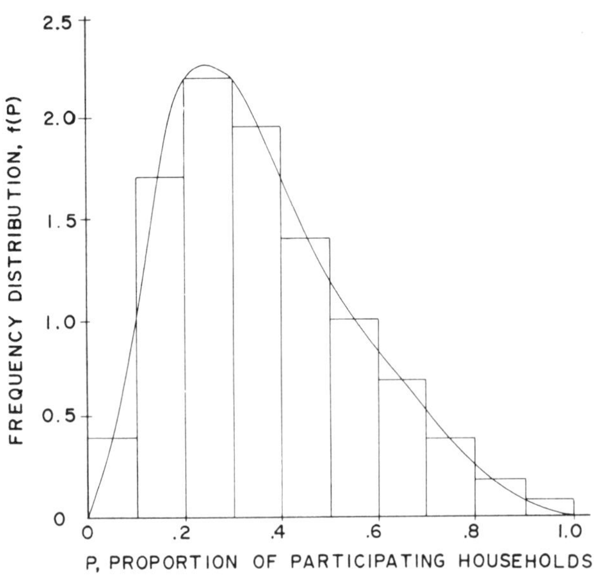

Figure 4-8. Construction of a frequency distribution from the cumulative probability curve of Figure 4-6.

Table 4-1. Construction of a Frequency Distribution from a Cumulative Probability Curve

(1) p Interval	(2) Interval Width	(3) F(p)	(4) Successive Differences of (3)	(5) Histogram Bar Heights (4)/(2)
0-0.1	0.1	0.04	0.04	0.4
0.1-0.2	0.1	0.21	0.17	1.7
0.2-0.3	0.1	0.43	0.22	2.2
0.3-0.4	0.1	0.62	0.19	1.9
0.4-0.5	0.1	0.76	0.14	1.4
0.5-0.6	0.1	0.86	0.10	1.0
0.6-0.7	0.1	0.93	0.07	0.7
0.7-0.8	0.1	0.97	0.04	0.4
0.8-0.9	0.1	0.99	0.02	0.4
0.9-1.0	0.1	1.00	0.01	0.1
			1.00	

4.4 THE COMBINATION OF EXPERT JUDGMENT: MATHEMATICAL APPROACHES

A problem may be important enough to justify consulting more than one expert. How this is the best use of these various expert judgments to be made? Whenever a number of individuals make assessments for the same situation, a consensus, which represents a combination of the individuals' assessments, can be formed. One approach is to combine the individual judgments by means of some sort of weighted average. There are at least three approaches to the problem of weights:

1. *Equal Weights*: In this case the decision-maker has no reason to think that there is much difference in ability among the experts, so he is willing to assign equal weights.
2. *Weights Based on Previous Performance*: This method may be subjective or objective, depending on the quantity and quality of the hard data obtained from prior experience with the use of experts.
3. *Weights Proportional to a Self-Rating*: Have each expert rate himself on a scale from 0 to M (conveniently M is either 10 or 100). Then assign a weight to each expert proportional to his self-rating, where the constant of proportionality is such that the weights sum to one (the constant is obviously the reciprocal of the sum of the self-ratings). The rationale behind this rule is that although a person may be able to make an assessment, the degree of belief in that assessment may vary, expert to expert. Also, although a person may be an expert in a given field, his expertise may vary from topic to topic within it.

Let us illustrate these rules with the example of Section 4.2 which, for the quadratic scoring rule (Equation 4.4), produced expected scores of 0.52 and 0.20, respectively, for reported estimated probabilities of 0.6/0.4 and 1/0 (where the first number represents the assessed probability of the first event and the other the assessed probability of the second event). We will assume now that these are the estimates produced by two *different* experts. If we weight the expected scores by the weights given by these three weighting rules, we obtain the weighted average of individual scores shown in the corresponding column of Table 4-2.

On the other hand, if we took a consensus of these two experts' probability estimates by averaging their probabilities according to the rule weights and then scored the consensus, we would obtain the results shown in the second scores column of Table 4-2. As an example of the calculation, consider Rule 1. Since $W_1 = W_2 = 1/2$, the averaged probabilities are 0.8/0.2 and the expected quadratic score is

$$0.6[2(0.8) - 0.64 - 0.04] + 0.4[2(0.2) - 0.64 - 0.04] = 0.44$$

What we notice is that the consensus scores are better than the weighted average of the individual scores.

Winkler [10] compared these three weighting rules using a betting situation involving hypothetical bets for NFL football games for the years 1966 and 1967. The scoring rules used included those of Equations 4-4, 4-5 and 4-6. With all three weighting rules, the consensus methods produced higher scores than the average of the individual subject's scores. The scoring rules are nonlinear functions, and it can be shown that the score obtained by a weighted average of probabilities must be greater than the respective weighted average of the corresponding individual scores. Over and above this characteristic of the scoring rules, however, the hypothetical bets made by Winkler's subjects showed that the consensus methods outperformed the average of the individuals. Although the consensus will always outperform the average score, it will not outperform *each* individual. There will always be some individuals who will outperform the consensus. Unless one is very confident that he can identify in advance the few experts who will "beat the consensus" in any given situation, however, he will be better off using a consensus.

Further, Winkler found little difference among the three weighting

Table 4-2. Comparison of Consensus and Nonconsensus Combinations

	Assumed Weights		Weighted Average of Individual Quadratic Scores	Consensus, Quadratic Scoring Method
	w_1	w_2		
Weighting Rule 1	1/2	1/2	0.36	0.44
Weighting Rule 2	3/10	7/10	0.30	0.36
Weighting Rule 3	2/3	1/3	0.41	0.49

consensus methods described above. Quoting Mosteller and Wallace from a different context, Winkler states: "In a way, this is an old story in statistics because modest changes in weights ordinarily change the output modestly."

4.5 THE COMBINATION OF PROBABILITY DISTRIBUTIONS: MATHEMATICAL APPROACHES

The preceding section concerned the combination of expert judgments when the judgments dealt with a number of discrete events. A judgmental probability distribution is quite another thing because the problem now becomes one of combining whole curves rather than simple probabilities. Assuming that two or more probability distributions could be weighted easily (this could be done by repeated random sampling experiments, but would require the use of a computer), nonconsensus averaging methods suffer from an additional fault in that a weighted average of unimodal distributions could result in a multimodal distribution. (This could result if the modes of the expert's probability distributions were substantially different from each other.) Thus, a paradox might occur in which the group distribution would be disavowed by each and every member of the group on the basis of shape alone. This suggests that a consensus method would be very desirable for the combination of individual probability distributions.

First it is desirable to find a distribution that provides a reasonable fit to the wide variety of frequency or cumulative probability distributions that may be encountered. Prior research has indicated that the underlying probability distributions obtained in practice tend to be lognormal in character [11]. This suggests that an individual's assessment of a probability distribution is determined by the product rather than the sum of a great many random factors. If so, then the logarithm of the probability distribution would then be the sum of independent factors. As the number of these random factors increases, the distribution of the logarithm of the probability distribution would approach a normal distribution according to the Central Limit Theorem of statistics, and the distribution would be lognormal. (Note: In fitting a lognormal distribution, the median *must* be greater than the mode.)

To fit a lognormal curve to a given probability distribution, the following relationships are helpful [12]:

$$X_{\text{median}} = \exp(m) \tag{4-7}$$

$$X_{\text{mode}} = \exp(m - v) \tag{4-8}$$

$$X_F = \exp[m + z\sqrt{v}] \tag{4-9}$$

$$f(X) = (1/X\sqrt{2v\pi})\exp\left[\frac{-(\ln X - m)^2}{2v}\right] \tag{4-10}$$

where m and $\sqrt{v}$ are the mean and the standard deviation of the normal distribution underlying the lognormal distribution, X_{median} and X_{mode} are two special values of the variable X, and X_F represents any value. (Since z is the standard normal deviate, according to Equation 4-9 every value of X corresponds to a particular value of z. For example, when X is at its median value, the value of z is zero for a normal distribution, so Equation 4-9 simplifies to Equation 4-7.)

Let us apply these formulas to fit a lognormal distribution to Figures 4-6 and 4-8 where we have labeled our variable p instead of X. From Figure 4-6, $F(0.50) = X_{median} = 0.34$. From Figure 4-8, $X_{mode} = 0.25$. Solving Equations 4-7 and 4-8 simultaneously, we find $m = -1.0788$ and $\sqrt{v} = 0.5545$. The calculations for f(X) are shown in Table 4-3 and are straightforward using Equation 4-10. The X-values are used as the midpoints of each histogram interval, and a smooth curve is drawn through the tops of the bars. The results are shown in Figure 4-9, as are the original data. As can be seen, the fit is very good.

To construct the fit of the cumulative probability distribution curve, we select convenient values of z and, from a table of the cumulative normal distribution, find the corresponding F(z) values (columns 3 and 4 of Table 4-3). Then Equation 4-9 is used to find values of the variable, X_F, corresponding to the z-value selected. Finally, F(z) is plotted against X_F to obtain the cumulative distribution curve. [Note that in Table 4-3, columns (1) and (2) are independent of columns (3), (4) and (5).] The results are shown in Figure 4-10, as is the original curve. Again, the fit is excellent.

At times, subjective probability distributions may appear to be normally rather than lognormally, distributed. If such is the case, then the distribution is easily fitted using the cumulative probability distribution, F(X). The

Table 4-3. Fitting a Lognormal Distribution

(1) X	(2) f(X)	(3) z	(4) F(z)	(5) X_F
0.05	0.04	-1.28	0.10	0.17
0.15	1.61	-1.00	0.16	0.20
0.25	2.47	-0.67	0.25	0.23
0.35	2.05	0.00	0.50	0.34
0.45	1.41	0.67	0.75	0.49
0.55	0.90	1.00	0.84	0.59
0.65	0.56	1.17	0.88	0.65
0.75	0.35	1.28	0.90	0.69
0.85	0.22	1.55	0.94	0.80
0.95	0.14	1.65	0.95	0.85

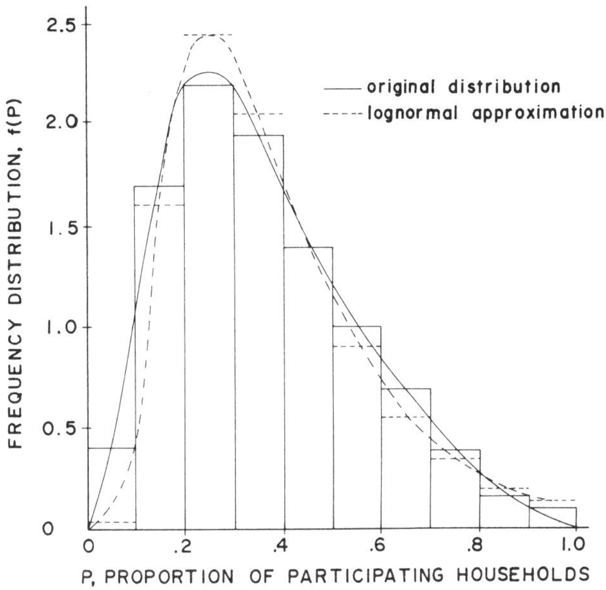

Figure 4-9. Fitting a lognormal distribution to a frequency distribution.

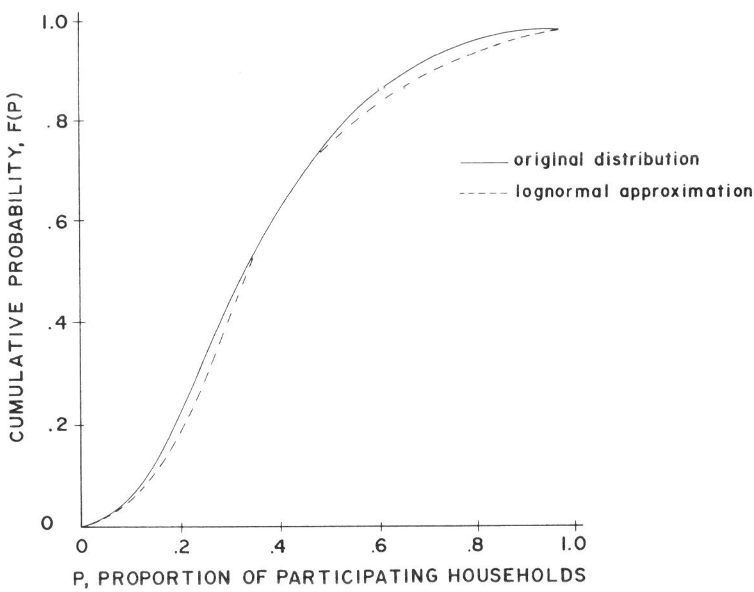

Figure 4-10. Fitting a lognormal distribution to a cumulative probability distribution.

F(X) = 0.50 point estimates the mean, m, and this together with the F(X) = 0.8413 point provides an estimate of the standard deviation $\sqrt{v}$ because

$$\sqrt{v} = X_{0.8413} - X_{0.50}. \tag{4-11}$$

Suppose F(110) = 0.50 and F(125) = 0.8413. Then m = 110 and $\sqrt{v}$ = 125 − 110 = 15. The fitted frequency and cumulative distributions can be plotted using the analogs to Equations 4-9 and 4-10:

$$X_F = m + z\sqrt{v} \tag{4-12}$$

$$F(X) = (1/\sqrt{2v\pi}) \exp[-(X - m)^2/2v]. \tag{4-13}$$

The calculations for f(x) and F(X) when m = 110 and $\sqrt{v}$ = 15 are shown in Table 4-4, and the curves are shown plotted in Figure 4-11. Again, as in Table 4-3, columns (1) and (2) are independent of columns (3), (4) and (5). The f(X) column can be obtained easily by dividing the f(X) values noted in the tables of the standard normal distribution $\sqrt{v}$. For the values of z selected in column (3), the F(z) values are also obtained directly from such tables.

Winkler [13] has proposed the use of Bayes' theorem to combine probability distributions from a number of experts. If f(X) is a prior frequency distribution and I represents the information in a sample, then Bayes' theorem can be written as follows:

$$f(X|I) \sim f(X)1(I|X) \tag{4-14}$$

where f(X|I) is the posterior frequency distribution (i.e., the distribution of X given I) and 1(I|X) is the likelihood of observing I given X. Statisticians have developed the concept of conjugate prior distributions, which essentially are families of distributions that ease the computational burden when used as frequency distributions. If the likelihood function is known, then the

Table 4-4. Fitting a Normal Distribution

(1) X	(2) f(X)	(3) z	(4) F(z)	(5) X_F
60	0.0001	−1.64	0.05	85.4
70	0.0008	−1.34	0.10	89.9
80	0.0036	−0.84	0.20	97.4
90	0.0110	−0.52	0.30	102.2
100	0.0212	−0.25	0.40	106.3
110	0.0266	0.00	0.50	110.0
120	0.0212	0.25	0.60	113.8
130	0.0110	0.52	0.70	117.8
140	0.0036	0.84	0.80	122.6
150	0.0008	1.34	0.90	130.1
160	0.0001	1.64	0.95	134.6

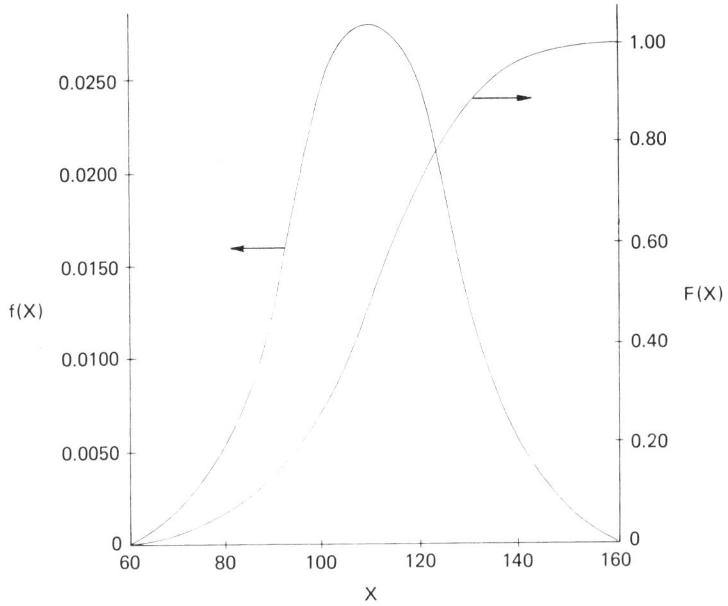

Figure 4-11. Fitting a normal distribution.

conjugate prior distribution results, using Equation 4-14, in the same kind of posterior distribution. Although the mathematical development is beyond the scope of the book, the results can be presented. For example, if the expert's judgmental probability distributions can be considered to be normal with means m_i and variances v_i (the subscript i signifying a particular expert), and if these individual normal distributions are weighted and combined via Bayes' theorem, the consensus would be a normal distribution with mean m and variance v as follows:

$$m = \sum(w_i m_i/v_i) \Big/ \sum(w_i/v_i) \tag{4-15}$$

$$v = 1/\Sigma(w_i/v_i) \tag{4-16}$$

where w_i is the weight applied to the ith expert.

In utilizing the conjugate method, although there is no theoretical restriction on the sum of the weights, if the judgments of k experts are completely independent in the sense that their judgments are based on independent sets of information (such as independent samples), then it seems reasonable to set each weight equal to one so that the sum of the weights equals k. On the other hand, if the judgments of the experts are based on exactly the same information (such as the same sample), the sum reasonably should equal one.

This suggests placing a lower bound on the sum of the weights of one, and an upper bound of k. Suppose that four experts have been asked to produce their probability distributions for the participating household problem of Figure 4-6, and that their median and modal values for p have been determined to be 0.34/0.25, 0.40/0.20, 0.45/0.30 and 0.30/0.25, respectively, where the first number in each pair is the median value and the second is the mode. Using Equations 4-7 and 4-8, the associated normal distribution parameters are found to be −1.0788/0.3075, −0.9163/0.6931, −0.7985/0.4055 and −1.2040/0.1823, respectively, where the first number in each pair is m_i and the second is v_i. We suppose that the four experts are knowledgeable about solid waste, in general, but that they specialize in different fields. Thus, their judgment could be considered to be somewhat, but not completely, independent so that we might set $\Sigma w_i = 2$. If we weight the first two experts twice as high as the last two, the weights are 2/3, 2/3, 1/3 and 1/3, respectively. Using Equations 4-15 and 4-16, m = −1.0515 and v = 0.1730. Using Equations 4-7 and 4-8, the parameters of the consensus distribution are found to be $X_{median} = 0.35$ and $X_{mode} = 0.29$. The complete consensus distributions, f(X) and F(X), can be calculated using Equations 4-9 and 4-10, as shown previously.

It is interesting to observe what occurs when the sum of the weights is varied holding the relative weights constant. The calculations just completed were repeated for k = 1 and k = 4, the results being shown in Table 4-5. Changing k does not alter the median of the consensus distribution. The quantity, v, i.e., the variance of the underlying normal distribution, however, decreases as k increases. In other words, as the degree of independence of the information on which the experts' judgments depend increases, the less variable is the consensus distribution. This is to be expected and is reflected also in the modal values that approach the median as k increases. The closer the mode is to the median in a lognormal distribution, the less variable is the lognormal distribution.

Table 4-5. The Effects of Varying k, the Sum of the Weights[a]

k	v	Median	Mode
1	0.3459	0.35	0.25
2	0.1730	0.35	0.29
4	0.0866	0.35	0.32

[a]For all values of k, the relative weights for the four experts are 1/3, 1/3, 1/6 and 1/6, respectively.

4.6 THE COMBINATION OF EXPERT JUDGMENT: BEHAVIORAL APPROACHES

Three techniques for dealing with the consensus problem may be described as behavioral in nature. The first, termed "Group Reassessment" (GR), involves both psychological factors and game-theoretic strategies. In brief, the experts meet as a group and discuss the matter, ultimately arriving at a group consensus. This is, of course, nothing more than the familiar "Committee," "Panel" or "Workshop" method. Psychological aspects involved include specious persuasion by the member with the greatest supposed authority or even merely the loudest voice; the unwillingness to abandon publicly expressed opinions; and the bandwagon effect of majority opinion. Studies have shown that the individual who is disproportionately influential in a group situation is extraverted, has a high need for achievement and is tolerant of ambiguity; his theoretical, economic and policital interests are high, as are his interpersonal relations [14]. It has been suggested [15] that group decisions are riskier than individual decisions, i.e., that individuals tend to shift their decision preferences after group discussion in the direction of greater risk-taking. However, recent research [16] has thrown doubt on the generality of the risky shift phenomenon, suggesting that it is situational-specific. Game-theoretic aspects must be considered also. For example, an expert may intentionally falsify his stated views in an attempt to influence the others. This is discussed in greater detail from another point of view in Section 4.2. The major advantage to GR is that it may lead to the consideration of factors that might otherwise be ignored.

As has been mentioned, the outcome of the GR method is apt to be a compromise among divergent views, arrived at all too often under the undue influence of certain psychological factors. One way to reduce these effects is to allow the experts to reassess their judgments in the absence of the other experts. This leads us to the second of the behavioral methods of arriving at a consensus, namely, Group Feedback and Individual Reassessment techniques (GFIR). Here, each expert separately reconsiders his assessments after being presented with feedback regarding all of the experts. This process of feedback and reassessment can be carried out repeatedly.

An interesting version of GFIR is the Delphi technique [9]. In general, Delphi procedures have three features: (1) anonymity—during a Delphi sequence the group members are not made known to each other; (2) controlled feedback—the exercise is conducted in a sequence of rounds between which a summary of the results of the previous round is communicated to the participants; and (3) statistical group response—the procedure presents a response that includes the opinions of the entire group summarized by measures of central tendency (means, medians, etc.) and variability (quantiles, variances, ranges, etc.). The most common form of Delphi is the paper and pencil version, in which a monitor team designs a questionnaire that is sent to the panel of experts (if the experts are scattered geographically, mail questionnaires are used). After the questionnaire is returned, the monitor team summarizes the results, which becomes the basis for a new questionnaire for the group. This constitutes a "round." The group members are given at least one opportunity to reevaluate their original answers based on examination of the group response.

As an example, each member of a resource recovery review panel was asked (individually) to examine a number of newly submitted proposals and to rate them according to a preestablished scale. The initial responses consisted of these ratings. A followup questionnaire fed back to the panel members included a summary of the distribution of these ratings in the form of the median for each proposal and—as an indication of the spread of opinions—the interquartile range (that is, the interval containing the middle 50% of the responses). Each panel member was then asked to reconsider his previous ratings and revise them if desired. If his new response lay outside the interquartile range for a particular proposal, he was asked to state his reason for thinking that the rating should be that much lower, or higher, than the majority judgment of the group. (In actual experience, placing the onus of justifying relatively extreme ratings on respondents has the effect of causing those without strong convictions to move their estimates closer to the median, while those who feel they have a good argument for a "deviationist" opinion tend to retain their original estimate and defend it.)

In the next round, responses (now usually spread over a smaller interval) were again summarized, and the panel members were given a concise summary of the reasons presented in support of extreme positions. They were then asked to revise their second-round responses, taking the proffered reasons into consideration and giving them whatever weight they thought was justified. A panel member whose answer still remained outside the interquartile range was required to state why he was unpersuaded by the opposing argument. In a fourth and final round, these criticisms of the reasons previously offered were resubmitted to the panel members, who were given a last chance to revise their ratings. The median of these final ratings was then to be taken as representing the nearest thing to a group consensus. Thus, four questionnaires and three summaries were used as follows:

1. Questionnaire I – Initial Response
2. Summary A – Means and Interquartile Ranges (M & IR)
3. Questionnaire II – Reconsideration and Statement of Reasons
4. Summary B – M & IR plus Summary of Reasons
5. Questionnaire III – Reconsideration and Reactions to Opposing Reasons
6. Summary C – M & IR plus Summary of Reactions to Opposing Reasons
7. Questionnaire IV – Final Response

The third behavioral approach is Concealed Feedback and Individual Reassessment (CFIR). This is similar to GFIR in that each expert separately reconsiders his assessments after being presented with some feedback regarding all of the experts. Unlike GFIR, however, the feedback is in the form of a reference point whose method of determination is not known to the expert. He is neither presented with a group consensus nor pressured to conform to a group consensus.

Ford [18] has described an innovative version of CFIR called "Shang inquiry." Whereas the Delphi technique draws its name from the oracle of Apollo and Delphi in ancient times, Shang owes its name to times even more ancient when the Shang rulers of China consulted oracular bones. Unlike the prophetic mutterings of the Delphic oracle, the oracular bones of the Shang produced only dichotomous answers, i.e., "yes or no," "favorable or unfavorable," etc. A typical Shang inquiry might go as follows, using the same example as was used to illustrate Delphi.

Round 1: For each proposal the experts are asked to make two estimates—one for the highest and one for the lowest plausible value that the proposal might rate on the preestablished scale. The group medians of these upper and lower boundaries are subsequently treated as initial boundaries for later rounds.

Round 2: The experts are presented with the midpoint of the boundaries established on the first round. It is given simply as a reference point, without indication as to how it was determined. Each expert responds with either "higher" or "lower" to indicate where he feels the true value falls relative to the reference point. In preparation for the next round, the responses are tabulated and, in accordance with the modal response, the group response (and presumably the "true" value) is located relative to the reference point. If the group is composed of an even number of experts, then ties are possible. The manager of the exercise may want to terminate the exercise with the tied reference point as the group response. But if it is desirable to continue, the exercise can proceed with the location of the group response being determined by a random process, e.g., toss a coin setting heads as higher and tails as lower. The exact position of that point is the midpoint of the new range with the last reference point as one boundary, the other boundary being either the upper or lower boundary from the previous range, depending on whether the group's modal response is "higher" or "lower." For example, if the modal response is "higher," then the previous upper boundary is maintained and the previous reference point becomes the new lower boundary. The midpoint of the range thus determined is treated as the group point estimate. Figure 4-12 is a flowchart of the procedure.

One consideration in the use of Shang inquiry is the number of iterations required to obtain a high level of accuracy where accuracy is defined as

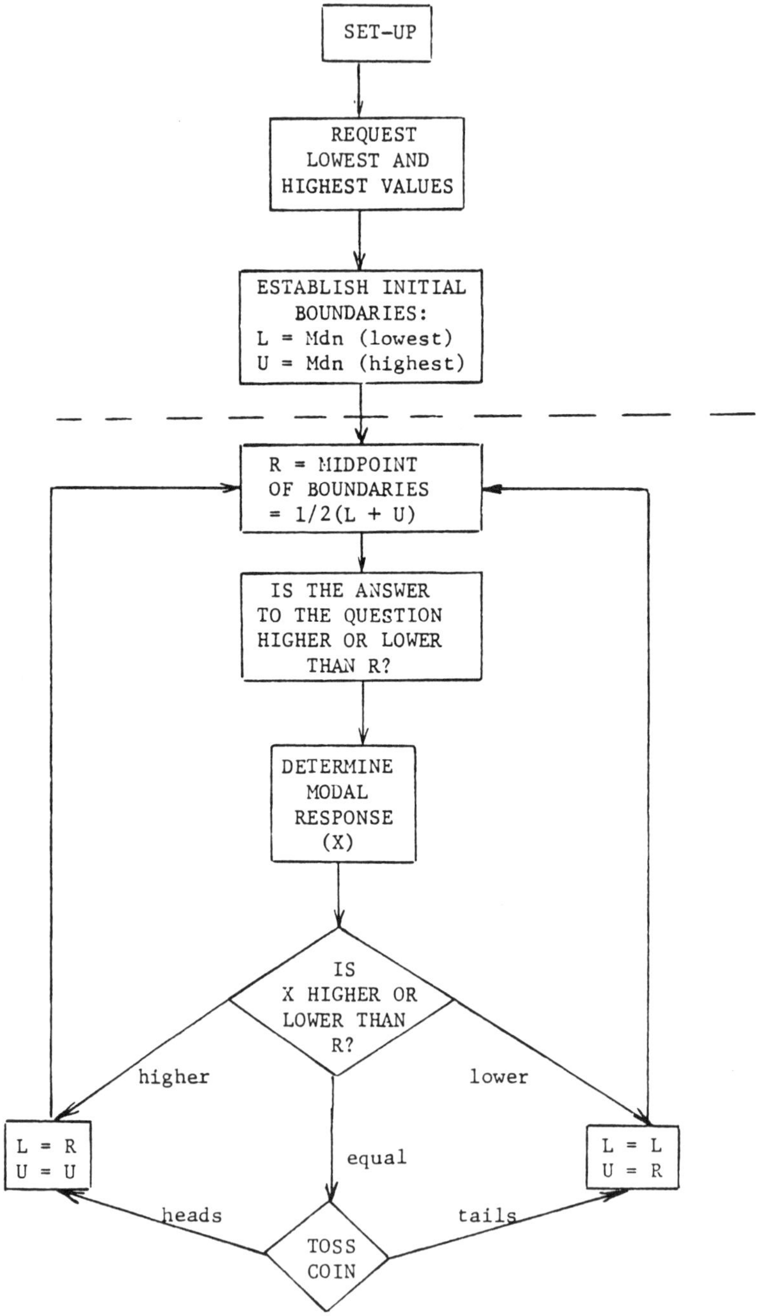

Figure 4-12. Flowchart of a Shang inquiry [17].

follows:

$$A_i = 1 - (D_i/D_{max}) \tag{4-17}$$

where D_i is the distance between the estimate and the true value in the ith iteration, D_{max} is the maximum possible error, and A_i is the accuracy on the ith iteration. For example, suppose the initial boundaries are Upper = 100, Lower = 0, and true value is 10. Further assume that the experts are perfectly accurate. The initial reference point is 50, and case a of Table 4-6 shows the accuracy achieved for six iterations. A worse case situation occurs when the true value is at the midpoint of the initial boundaries (case b in Table 4-6). Even so, more than 98% accuracy is achieved at the end of six iterations.

It should be noted that the techniques described in Section 4.3 of eliciting single probabilities or a probability distribution can be applied with behavioral as well as with mathematical approaches. With GR techniques, the training described in Section 4.3 can be done before the beginning of group deliberations. With GFIR and CFIR techniques, the training can also be done as a group (because the training does not involve group interaction) or individually.

4.7 A COMPARISON OF CONSENSUS TECHNIQUES

Much has been written about Group Reassessment [19] and considerable research has been conducted in the area. Given the nature of the subject, it is easy to see why it is difficult to draw firm conclusions from the experimental findings, but the generalizations they suggest can be mentioned briefly.

1. *Group Performance*: Group performance depends on the characteristics of the problem undertaken. For the simpler type of problem, group performance is generally at the level of the most proficient member. Group performance is above the level of the most proficient member when the problem consists of multiple parts and when information is not uniformly held among the members. Finally, group performance can be below the level of the most proficient member when the task involves starting at several different points in a problem but all following some sort of consistent strategy. If the group members do not follow a common strategy, they create "noise" for one another. The verbalizations of several experts who have started at different points and are pursuing different lines of reasoning may be mutually disruptive.

2. *Consensus*: Group problem discussion generates pressures toward uniformity.

3. *Speed*: Groups are slow compared with individual action.

4. *Coordination*: On the whole, groups are inferior to individuals in tasks requiring coordination of efforts.

Table 4-6. Accuracy in Shang Iterations

Iteration	Reference Point	Error	Accuracy
	(a) True Value = 10		
1	25	15	0.833[a]
2	12.5	2.5	0.972
3	6.25	3.75	0.958
4	9.375	0.625	0.993
5	10.938	0.938	0.990
6	10.157	0.157	0.998
	(b) True Value = 50		
1	25 (or 75)	25	0.500
2	37.5 (or 62.5)	12.5	0.750
3	43.75 (or 56.25)	6.25	0.875
4	46.88 (or 53.13)	3.13	0.938
5	48.44 (or 51.56)	1.56	0.969
6	49.22 (or 50.78)	0.78	0.984

[a]Example of typical calculation: The estimate is 25 and the true value is 10. Therefore, $D_i = 15$. The maximum possible error is $100 - 10 = 90 = D_{max}$. From Equation 4-17, $A_i = 1 - 15/90 = 0.833$.

5. *Organization*: Groups require organization and time to develop it.

6. *Risk*: As mentioned before, under some circumstances group decisions are riskier than individual decisions.

7. *Commitment*: With regard to the effects of groups on the motivation of their members, commitment to the group decision is generally higher than to the individual decision. The effect is sensitive to group conditions, however. External planning and organization for the group lessens such commitment.

There is also a considerable body of research and literature on GFIR, particularly the Delphi technique [17] and, if anything, there is more controversy over GFIR than over GR. The following are some generalizations of the results of the research conducted.

1. *Delphi Performance*: The validity of Delphi has generally been measured by use of "almanac" questions, i.e., where there was a verifiable answer to check the performance of individuals and groups and where the subjects did not know the answer but had sufficient background information so they could make an informed estimate. A typical example is that shown in Table 4-7 [20]. As the original questionnaire involved no feedback, the median value represents the consensus that would be obtained using the mathematical approaches outlined in Sections 4.5 and 4.6 (the median of a lognormal

Table 4-7. Results of a Delphi Exercise [20]

Question	Results of Original Questionnaire			Results of First Iteration			Correct Answer
	$X_{0.25}$	X_{median}	$X_{0.75}$	$X_{0.25}$	X_{median}	$X_{0.75}$	
1. What was the number of telephones (in thousands) in Asia in 1967?	150	5,000	1,000,000	500	3,000	10,000	21,758
2. How many million barrels of beer were produced in the U.S. in 1966?	3	50	240	20	77	240	110
3. How many millions of dollars were expended for public elementary and secondary education in the U.S. in 1967?	100	982	10,000	500	5,000	8,000	31,511
4. What was the total U.S. Budget income in 1968 in billions of dollars?	10	200	500	100	200	300	165
5. How many thousands new private housing units were started in the U.S. in 1969?	80	500	1,000	100	500	640	1,490
6. In 1967 what was the average number of pounds of milk consumption per person in the U.S.?	50	200	700	52	200	500	286
7. What was the total motor fuel consumption in millions of gallons in 1967 in New York State?	200	1,000	6,000	500	1,000	5,000	5,057
8. What was the total amount of deaths from falls in the U.S. in 1967?	1 000	5,000	25,000	1,000	8,000	10,000	19,800
9. How many elementary and secondary school teachers were there in the U.S. in 1967?	400 000	800,000	1,000,000	800,000	1,000,000	1,000,000	1,854,700
10. How many radio and TV broadcast stations were operating in the U.S. in 1967?	1,000	5,000	10,000	2,500	5,000	6,000	6,253

consensus distribution is independent of the weighting of the individual distributions, as was shown in Section 4.6). As Table 4-6 shows, the Delphi approach produced median estimates superior to the individual approach four times (questions 2, 3, 8 and 9), identical results five times (questions 4, 5, 6, 7 and 10), and a less accurate result in only one case (question 1). Using the difference between the correct answer and the percentile farthest from the original questionnaire as the maximum error, the average accuracy of the individual method can be shown to be 0.446 using Equation 4-17. For the Delphi results, the average accuracy is 0.509. Other data are available that suggest that the accuracy obtained through Delphi is also superior to GR techniques as well, at least with respect to almanac-type questions.

2. *Consensus*: As can be seen from Table 4-6, the spread of the Delphi answers decreased in about all cases from the original individual responses. Generally speaking, although there is a reduction in the variance of the estimates as the number of rounds is increased, most of the reduction is obtained by the third iteration.

3. *Organization*: Delphi depends on formal questionnaires to elicit responses from the experts used in the study. The questionnaire must be carefully designed to provide unambiguous stimuli to the experts to avoid bias or distorted responses. Further, it has been shown [21] that deliberate manipulation of Delphi feedback can significantly shift the responses.

4. *Group Size*: Delphi performance generally increases with the number of experts employed, up to a total of between 5 and 11.

There is a lack of research with regard to CFIR, but the following generalizations are based on the work of Ford [18].

1. *Shang Performance*: Compared to Delphi and individual performance, Shang was superior to both with regard to accuracy. Further, there was more change induced from the initial to the final estimates than either with Delphi or individual assessment. This is not surprising because change is almost certain, due to the nature of the group statistic construction in Shang. Other things being equal, if an individual changes his position, he is unlikely to violate group opinion. A median can only change if individual response shifts either cross the median or, if they are initially at the median, move away from it. Apparently, the Delphi method includes change toward the median but rarely causes the median to change.

2. *Difficulty For Experts*: Experts find the questions posed by Shang inquiry more difficult to answer than those of either the individual or Delphi questionnaires. Most subjects found they had to rethink a Shang question on every iteration, in part because they did not have the security of a group reference point.

Table 4-8 summarizes the major consensus techniques, including mathematical combination (MC). As can be seen, there is no one technique that dominates the others. MC rates high on 7 of the 10 factors in Table 4-8 and low on 3 of them, but its primary weakness is that it lacks feedback from the other experts, a fault not found in any of the behavioral methods. It also rates low on commitment to group consensus, something that might be an important factor with regard to "selling" or otherwise implementing a decision.

The primary advantage of GR is the possibility of introducing new factors or considerations not previously thought of by the decision-maker. Its primary disadvantage, of course, lies in the adverse social pressures present during group interaction. There is high commitment to group consensus, however, which is a valuable positive aspect.

As with GR, GFIR rates high on about half of the 10 factors but presents a vastly different profile because GR and GFIR are similar only with respect to feedback and commitment. It is an improvement over GR in that the adverse social pressures are absent. Because of the written explanation of extreme positions, there is some opportunity for the introduction of new factors in GFIR.

CFIR resembles GFIR on 7 of the 10 factors and, like GFIR, rates high on half of them. Although it loses the commitment to a group consensus, by this

Table 4-8: Summary of Consensus Techniques

	MC	GR	GFIR	CFIR
Can be implemented via mail questionnaire	Yes	No	Yes	Yes
Pressure to arrive at consensus	No	Yes	Yes	No
Presence of other adverse social pressures	No	Yes	No	No
Consensus can be manipulated by administrator	No	No	Yes	Yes
Degree of difficulty for administrator	Low	Low	High	Medium
Degree of difficulty for experts	Low	Low to high	Medium	Medium to high
Incorporates feedback	No	Yes	Yes	Yes
Experts can be independently weighted	Yes	No	Yes	Yes
Introduction of new factors previously unthought of by administrator	No	Yes	Partially	No
Commitment to group consensus	Low	High	High	Low

very fact it minimizes the pressure for group consensus. This is clearly an advantage because if the experts in a Delphi exercise feel that the questionnaire is an imposition on them or that they do not have time to give adequate thought to the questions, they will agree with the majority simply to avoid having to explain the differences. In this respect, therefore, the Delphi procedure is not an absolute guarantee against the "bandwagon" effect. It should be noted that the Shang technique does require twice the number of iterations to reduce error as does Delphi, but the author feels that compared to Delphi, the only really important disadvantages are the lack of commitment to group consensus and the absence of a mechanism for the introduction of new factors (which Delphi has only in a restricted sense). In most other regards, Shang is superior to Delphi and is preferred by the author.

4.8 SUMMARY

There are two options available when one is working on a problem under conditions of uncertainty with insufficient data, incomplete theory or a high order of complexity:

1. We can wait indefinitely until we have adequate data and theory to enable us to deal with such problems as confidentally as we do with problems in physics or chemistry.
2. We can make the most of an admittedly unsatisfactory situation by trying to obtain the relevant intuitive insights of experts and using their judgments as systematically as possible.

This chapter has dealt with the second of these two options.

However, expert opinion is frequently diverse as well as diffuse, and the decision-maker may find it difficult to employ a multitude of estimates when he is really only looking for one. In such cases, either some sort of average or consensus must be obtained as input to the next stage of decision-making. Both mathematical and behavioral approaches have been outlined and compared in this chapter, but since the choice of approach depends on the relative importance of the basic characteristics of these techniques (i.e., their advantages and disadvantages), the final selection must be made by the decision-maker himself.

REFERENCES

1. Helmer, O., and N. Rescher. "On The Epistemology of the Inexact Sciences," *Management Sci.* 6(1):25-52 (1959).

2. Dalkey, N. C. "Toward a Theory of Group Estimation," in H. A. Linstone and M. Turoff, Eds. *The Delphi Method: Techniques and Applications* (Reading, MA: Addison-Wesley Publishing Co., Inc., 1975), pp. 236-261.
3. Asch, S. E. "Effects of Group Pressure Upon the Modifications and Distortion of Judgment," in *Readings in Social Psychology*, E. E. Maccoly, T. M. Newcomb and E. L. Hartley, Eds. (New York: Henry Holt, 1951), pp. 174-183.
4. Wallich, H. C. "The American Council of Economics Advisors and the German Sachverständigenrat: A Study on the Economics of Advice," *Quart. J. Econ.* 82:350-379 (1968).
5. Winkler, R. L. "The Quantification of Judgment: Some Methodological Suggestions," *J. Am. Stat. Assoc.* 62(320):1105-1120 (1967).
6. Savage, L. J. "Elicitation of Personal Probabilities and Expectations," *J. Am. Stat. Assoc.* 66(336):783-801 (1971).
7. Wilde, D. J. "Single Variable Search," in *Optimum Seeking Methods* (Englewood Cliffs, NJ: Prentice-Hall, Inc., 1964), pp. 23-24.
8. Raiffa, H. "More about the Economics of Sampling," in *Decision Analysis: Introductory Lectures on Choices Under Uncertainty* (Reading, MA: Addison-Wesley Publishing Co., Inc., 1970), pp. 161-168.
9. Winkler, R. L. "The Assessment of Prior Distributions in Bayesian Analysis," *J. Am. Stat. Assoc.* 62(319):776-800 (1967).
10. Winkler, R. L. "Probabilistic Prediction; Some Experimental Results," *J. Am. Stat. Assoc.* 66(336):675-685 (1971).
11. Martino, J. P. "The Lognormality of Delphi Estimates," *Technol. Forecasting* 1:355-358 (1970).
12. Aitchison, J., and J. A. C. Brown. "General Properties of Lognormal Distributions," *The Lognormal Distribution* (New York: Cambridge University Press, 1976), pp. 7-19.
13. Winkler, R. L. "The Consensus of Subjective Probability Distributions," *Management Sci.* 15(2):61-75 (1968).
14. Rim, Y. "Who Are the Risk-Takers in Decision-Making?" *Personnel Admin.* 29:26-30 (1966).
15. Vinokur, A. "Review and Theoretical Analysis of the Effects of Group Processes Upon Individual and Group Decisions Involving Risk," *Psychol. Bull.* 76:231-250 (1971).
16. Felsenthal, D. S. "Group Versus Individual Gambling Behavior: Reexamination and Limitation," *Behavioral Sci.* 24(5):334-345 (1979).
17. Pill, J. "The Delphi Method: Substance, Context, A Critique and an Annotated Bibliography," *Socio-Econ. Planning Sci.* 5:57-71 (1971).
18. Ford, D. A. "Shang Inquiry as an Alternative to Delphi: Some Experimental Findings," *Technol. Forecasting Social Change* 7:139-164 (1975).
19. Kelly, H. H., and J. W. Thibaut. "Group Problem Solving," in *Handbook of Social Psychology*, Vol. IV, G. Lindzey and E. Aronson, Eds. (Reading, MA: Addison-Wesley Publishing Co., Inc., 1969), pp. 1-101.
20. Doyon, L. R., T. V. Sheehan and H. I. Zagor. "Classroom Exercises in Applying the Delphi Method for Decision-Making," *Socioecon. Planning Sci.* 5(4):363-375 (1971).
21. Nelson, B. W. "Statistical Manipulation of Delphi Statements: Its Success and Effects on Convergence and Stability," *Technol. Forecasting Social Change* 12:41-60 (1978).

CHAPTER 5

FORECASTING

5.1 INTRODUCTION

Forecasting is an attempt to estimate or calculate something measureable in advance or to predict whether a given event will or will not occur. For example, in the design of a resource recovery plant, the former might involve the estimation of the quantity of municipal solid waste that will be generated in a given area a decade from now. The latter might involve the prediction of whether a specific piece of legislation will be passed that would affect the solid waste quantity generated. These examples illustrate two rather basic types of entities encountered in forecasting for solid waste resource recovery purposes.

Table 5-1 lists several forecasting techniques of interest in resource recovery planning. It is not an exhaustive list because it omits many methodologies that, in the opinion of the author have little application to the problem at hand. An example of an omitted methodology is the Precursive Indicator, i.e., a forecast that can be developed by correlating trends between two developments, one of which is leading the other. This approach is useful as long as the leader–follower relationship is maintained, but for the long-term forecasts that concern us here there is as much uncertainty in forecasting the leader as forecasting the follower. Also not included in Table 5-1 are methodologies based on mixtures of these techniques, such as the Abt model, which incorporates scenarios; applies probabilities of attaining the various futures; modifies the results by intuitive expert short-range feedback; and integrates these steps by computer simulation. Excellent expositions of a number of other techniques are provided by Jantsch [1] and Lanford [2], should the reader desire a more comprehensive listing of forecast methodologies.

Short definitions of each of the methodologies listed in Table 5-1 are as follows:

1. *Individual*: a forecast made by a single individual who is expert in his area of specialty and knowledgeable in related scientific disciplines or terminologies.
2. *Poll*: a forecast made by combining the judgments of several individuals who are active and, presumably, expert in the field.
3. *Panel/Committee*: a forecast resulting from fact-to-face interaction within a group of individual experts.
4. *Delphi*: a forecast made as the result of the systematic solicitation of expert opinion working toward a consensus by use of sequential individual written interrogations interspersed with information and opinion feedback derived from earlier parts of the program.
5. *Scenarios*: a projection of the future according to present conditions, with results of stated assumptions taken into account. Assumptions include economic, political, social and technical possibilities. Scenarios depict possible alternative futures, depending on the assumptions.
6. *Trend Extrapolation*: a forecast based on the development of an historic time-series, holding the assumption that the factors contributing to the trend of the historical series in the past are more likely to remain constant than to change in the future time period being considered, and that the extrapolation of this trend line of time-series into the future represents the path of future progress of the subject area under discussion. The most

Table 5-1. Selected Forecasting Methodologies with Application to Resource Recovery Planning and Design

Individual
Polls
Panels/Committees
Delphi
Scenarios
Trend Extrapolation
Straight line
Nonlinear
Analogies
Biological growth—Pearl's law
Economic growth—Gompertz' law
Network Construction
Decision trees
Cross-impact analysis

often encountered trend extrapolation technique involves straight-line and fitted curve extrapolation where the extrapolation is of future points from the trend line of historical data.

7. *Analogies*: a forecast made by analogy to growth in other systems. One well-known example is analogy to biological growth (Pearl's law) where the initial advance is first exponential, then symmetrical about an inflection point, followed by a continued diminution of the rate of progress as it approaches an upper limit set by known physical limits or other positive and unalterable limits known to the forecaster. Another involves an analogy to economic growth (Gompertz' law) where the initial advance is exponential (but where the curve is not symmetrical about an inflection point), followed by a continuous diminution of the rate of progress as it approaches a perceived physical limit.

8. *Network construction*: a structuring device that depicts, graphically, the relation of pertinent background factors, levels of use, functional possibilities, expected effort and paths of development to stated objectives. The most important of the network construction methods is the decision tree approach, which uses a series of nodes and branches to indicate the interrelationship of action choices with different possible events or results of action that are partially affected by chance or other uncontrollable circumstances. The decision tree method can vary from a simple finite branching network to a stochastic network in which formal probabilistic methods are used. Another important network method is the cross-impact matrix, which attempts to organize and process information about interrelationships in a way that minimizes these limitations, while evaluating the permutations and combinations of the events considered. A cross-impact analysis begins with the derivation of a list of forecasts concerning these developments. The potential interactions between members of each pair of developments on the list are assessed. (This requires answering questions such as "If the first development on the list were to occur, how would the likelihood of occurrence of the second development be affected?") The synthesis of these judgments results in revisions of the estimates of the probability of occurrence of the developments on the list. This technique is useful in evaluating the effects of combinations of many potential occurrences.

The first five methodologies (individual, poll, panel/committee, Delphi and scenarios) generally are concerned with the utilization of expert judgment (UEJ) and are often incorporated as part of another forecast methodology to provide input. Most of these methods are discussed in detail in Chapter 4. The decision tree approach is described in Chapter 3. The analogy method is often paired with utilization of expert judgment techniques because the latter supplies input to the model suggested by the former.

5.2 TREND EXTRAPOLATION

Because it is so widely used, it is important to consider trend extrapolation in greater detail. Such methods are generally concerned with the estimation of measurable qualities, and they are primarily statistical in nature. They incorporate the concept that an observation in time has two components: (1) the true value or deterministic component, and (2) a random or stochastic component. Based on historical data, such models are used to calculate the probability of a future value lying between two specified limits.

The basic premise of these models is that what will happen in the future is the result of factors already present and operating today. One of the simplest forecasting models of this type is the linear trend of the following form:

$$Y = B_0 + B_1X + e \tag{5-1}$$

where Y is the variable of interest (quantity of solid waste, for example), X is the time axis, B_0 and B_1 are the parameters (constants) of the model, and e is a random disturbance. Typically, the parameters are estimated by regression techniques, and the standard error, S_e, of an individual forecast is given as [3]

$$S_e = S_d \sqrt{1 + 1/N + x^2/\Sigma x_i^2} \tag{5-2}$$

where S_d is the standard error of the regression, N is the number of observations in the data base, x is the number of time units (from the mean of the data base) at which we wish to predict Y, and Σx_i^2 is the sum of the squared time data base points about the mean. For any given data base, S_d, N and Σx_i^2 are constants.

Assuming for the moment that S_d is measured without error (so that normal distribution theory can be used), a 90% confidence interval of an individual forecast about the trend line is shown by the curved lines in Figure 5-1. At x = 0 (i.e., the midpoint of the data base), S_e is at its minimum, as is the 90% confidence interval (for large N) of an individual forecast, i.e., $\pm 1.645 \sqrt{1 + 1/N}$. This minimum is shown by the horizontal lines. (Not knowing S_d, the ordinate in Figure 5-1 is scaled in units of $S_d \sqrt{1 + 1/N}$.) Thus, at x = 0, we are 90% confident that our individual forecast lies within the horizontal lines, but as we move away from the data base in time, the 90% confidence limits widen so that we are less confident that the true quantity lies within the original minimum confidence interval. In effect, this represents a loss of confidence (in the statistical sense) with time.

Fortunately, the new confidence at a given forecast time can be calculated rather simply [4]. The basic procedure is to compute an adjusted t-value, t′:

$$t' = \frac{t\sqrt{1 + 1/N}}{\sqrt{1 + 1/N + x^2/\Sigma x^2}} \tag{5-3}$$

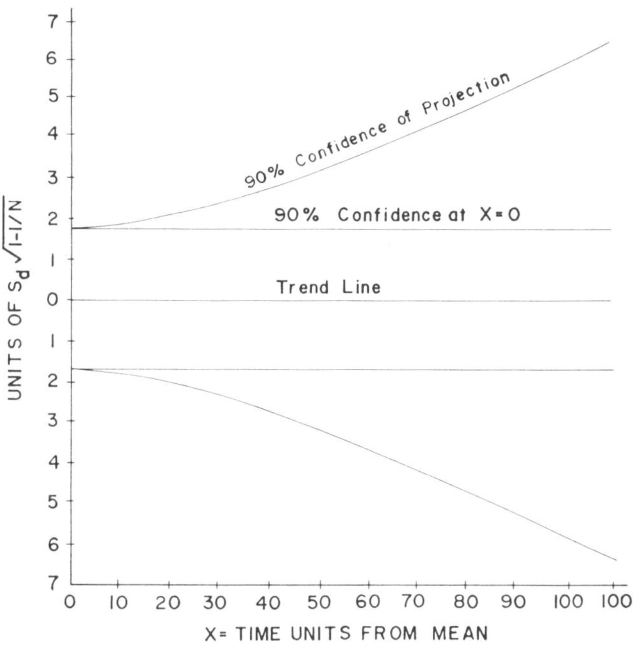

Figure 5-1. A 90% confidence interval for an individual forecast.

where t is the Student t-value at N − 2 degrees of freedom (df) and the selected significance level. Tables of the t-distribution are used then to find the new confidence level.

For example, suppose we wish to predict quantity at 10 years (120 months) beyond the end of a 24-month data base (i.e., observations taken at monthly intervals). Then

$$N = 24$$

$$\bar{X} = 12.5,\ \Sigma X = 300 \text{ and } \Sigma X^2 = 4900$$

$$x = 144 - 12.5 = 131.5$$

$$\Sigma x^2 = 4900 - (300)^2/24 = 1150$$

$$\sqrt{1 + 1/N} = 1.0206$$

$$\sqrt{1 + 1/N + x^2/\Sigma x^2} = 4.0098$$

$$t = 1.717 \text{ (df = N - 2 = 22, 90\% confidence level)}$$

Thus, t′ = 1.717(1.0206)/4.0098 = 0.437. Interpolating in the Fisher and Yates t-tables [5], we find the confidence equal to 33%. In other words, our forecast at 10 years beyond the end of the data base has only a 33% chance of falling within the original confidence interval. Repeating the calculations, at 20 years beyond the end of the data base (20 years is an average depreciation period for a resource recovery facility), this confidence drops to 18%. The weakness of the statistical forecast is clear, i.e., the errors of the forecast are exceedingly large even when we assume that this rather simple model is valid. This approach, however, is valuable in that it establishes an objective standard rate of loss of confidence against which subjective judgment can be compared. This is explained in the next section.

5.3 ESTIMATING FUTURE CONFIDENCE

A 99% confidence interval would include most of the observations we would expect to see if forecasting using the linear trend model of Equation 5-1. Applying Equation 5-3 with t at the corresponding significance level of 0.01, the curves shown in Figure 5-2 were generated, showing the deterioration of the starting 99% confidence with time. (The greater the number of observations in the data base, the slower is the deterioration.) Using the linear trend model of Equation 5-1 as an objective standard then, we have a method

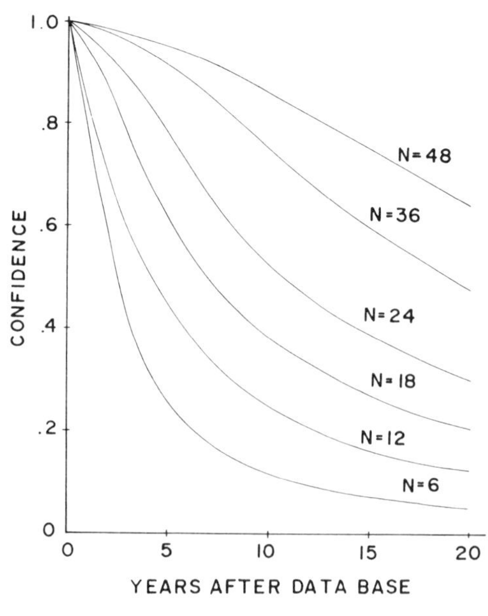

Figure 5-2. Deterioration of confidence with time.

by which we can weight the value of future beliefs, which are based either on an actual data base or a current belief.

If the current estimate of an effect is based on an actual data base, the particular curve to be selected from Figure 5-2 is determined by N, the number of observations in the data base. (The curves in Figure 5-2 were generated assuming that the data were collected at monthly intervals. If the actual data were taken over different intervals, the reader can generate his own curves using Equation 5-3.) If, on the other hand, the current estimate of an effect is based on a subjective judgment, the hypothetical sample method can be used [6]. In the hypothetical sample method, the estimator is saying, "This is the estimate I believe we would obtain if we had N actual data points." In other words, although no actual sample has been obtained, the estimator considers his judgment to be roughly equivalent to the information from such a sample. The size of N reflects the strength of his belief in his current estimate. If he is unsure of himself, his belief would correspond to a small sample size with all its uncertainties; if his belief is strongly held, he would consider it equivalent to a large actual sample.

Since Figure 5-2 is based on an objective model, one can use its results as an upper bound on the future confidence of subjective assessments. One can use Figure 5-2 without further modification or, alternatively, it can be used as a starting point in some selected UEJ exercise.

5.4 GROWTH CURVES

The use of analogy is not only a common and natural way of thinking, it is of value to empirical science because the development of theory is enhanced by the recognition of patterns that ultimately lead to higher levels of abstraction [7]. In this sequence, the formulation of a hypothesis roughly follows pattern identification; the creation of a theory follows hypothesis testing; and prediction follows theory. The problem with forecasting lies in the hypothesis testing stage. One usually cannot afford the wait to compare the actual to the forecasted results because the forecast is often used for decisions that must be made in the present. If we assume, however, that there are similarities among two or more entities and that these entities are independent of each other in time, then the presumption is that we can evaluate one by analogy to the other.

Of course, analogies can be spurious even though they might appear to be objective. The use of growth curves lends the illusion of rigor simply because of their mathematical nature. On the other hand, a growth curve might be selected on the basis of the requirements of a forecast model specified by expert judgment. In such an instance it is not the analogy that is the basis for our forecast; the analogy is used because it meets predetermined specifications. As an example, suppose we are interested in forecasting one or more system variables (e.g., quantity of solid waste, composition, etc.). Then the following properties might be deemed desirable in a model:

1. The system variables are bounded. Since any variable of human significance cannot increase indefinitely, there must be distinct limits.
2. A variable increases or decreases according to whether the net impact of the other variables is positive or negative.
3. A variable's response to a given impact decreases to zero as that variable approaches its upper or lower bound.
4. All other things being equal, a variable will produce a greater impact on the system as it grows larger.

If it were agreed that the above would be requirements of a given forecast model, the next step would be to find such a model. It would quickly be determined, however, that although many models would be unacceptable according to these criteria (e.g., the exponential function because growth continues infinitely in that model), a large number of models would qualify. All other things being equal, there is no reason why the simplest model should not be chosen (Ockham's rule or the Law of Parsimony.)

Such a model is the Pearl Curve, named after the American biologist who made extensive studies of the growth of organisms and populations and found that the model described their growth very well. The Pearl Curve is often referred to as the logistic curve. In physics, it is identical to the Fermi-Dirac distribution. Its form is as follows:

$$y = \frac{L}{1 + ae^{-bt}} \tag{5-4}$$

where L is an upper limit to growth of the variable represented by y, and a and b are parameters (constants) of the model. An example of the Pearl Curve is shown in Figure 5-3. The curve has an initial value of zero at t equal to minus infinity, and reaches the limiting value, L, at time, t, equal to plus infinity. The curve is symmetrical about an inflection point found at t equal to zero. As it turns out, a number of phenomena display S-shaped growth curves similar to the Pearl Curve. However, we are not accepting the Pearl Curve here *because* of such similarities, but rather because it is one of the simplest of the set of models that satisfy our predetermined requirements.

5.5 A FORECASTING METHODOLOGY

In addition to the large errors experienced in forecasting far beyond the data base when traditional statistical models are employed, their basic assumption that no new or important events will occur throughout the forecast period is clearly untenable. A trend that has persisted for hundreds

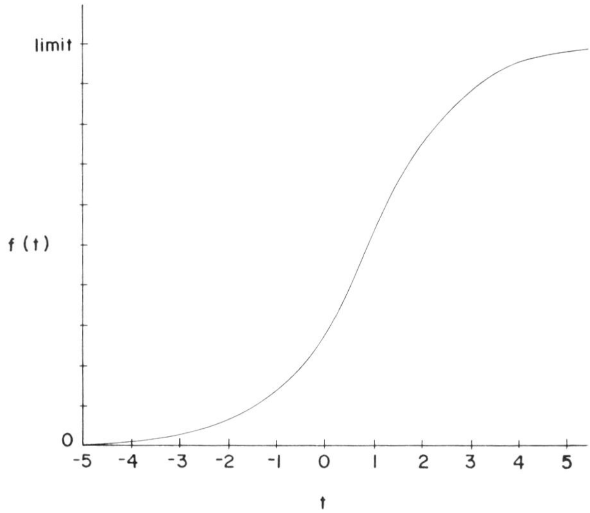

Figure 5-3. Example of a Pearl Curve (logistic function).

or even thousands of years may change within a decade or even more rapidly. Further, in the real world things do not grow until they reach infinity; rather, growth eventually reaches some limit at which the process must slacken.

Incorporating selective features from many of the methodologies listed in Table 5-1, the author has sought to devise a solid waste forecast model that (1) would be capable of dealing with both dichotomous events and measurable variables; (2) would not involve a complicated structure; (3) would not require computer simulation for its solution; and (4) would meet the four requirements outlined in the previous section. Such a model is based on the work of Kane and his colleagues [8,9] and is expressed in two equations:

$$f(t+1) = f(t)^{p(t)} \tag{5-5}$$

where f(t) is the value of the variable we are trying to forecast at time t, f(t + 1) is the value of the variable at one time period afterward (a time period can be a week, a month, a year, etc.), and p(t) is given by the following:

$$p(t) = \frac{1 + k/2\ \Sigma(|c_j| - c_j)x_j}{1 + k/2\ \Sigma(|c_j| + c_j)x_j} \tag{5-6}$$

where x_j is the value of the jth variable at time t, c_j is the impact factor at time t of variable j on the variable to be forecast, k is the reciprocal of the maximum number of time periods to be forecast, and the limits on the summation signs are j = 1, 2, . . . , N, where N is the number of variables affecting the variable to be forecast. Equation 5-6 can be rewritten as follows:

$$p(t) = \frac{1 + k \; \Sigma \, |(\text{negative impact factors}) \; x_j|}{1 + k \; \Sigma \, |(\text{positive impact factors}) \; x_j|} \tag{5-7}$$

and this is the form actually used in the methodology to be described.

When the weighted negative impacts are greater than the positive ones, $p(t) > 1$ and $f(t + 1)$ decreases; when the weighted positive impacts are less than the positive ones, $p(t) < 1$ and $f(t + 1)$ increases. When they are equal, $p(t) = 1$ and $f(t + 1)$ remains the same. Thus, condition 2 of the previous section holds. By rescaling any factor 0-1 (which we can do since all practical variables are bounded), condition 1 is satisfied. Although it will not be shown here, for small increments in time Equations 5-5 and 5-6 are solutions to the following differential equation:

$$\frac{df}{dt} = \Sigma a_j x_j x_i \ln x_i \tag{5-8}$$

where x_i is the forecast variable at time t. When x_i approaches 0 or 1, then df/dt approaches 0, which is condition 3. Considering the other variables (x_j) individually, as an x_j increases or decreases so does its impact on the variable forecasted. This is condition 4.

5.6 AN EXAMPLE OF THE MODEL–INITIAL DATA AND ASSUMPTIONS

As an example, we will consider forecasting the quantity of solid waste generated in a given waste shed over a period of 10 years.

A. *Initial Data*
   1. The current generation rate is 3.5 lb/person/day.
   2. There is an historical trend of an increase of 2%/yr.

B. *Assumptions For The Future*
   1. The maximum value the generation rate can attain over the next 10 years is 5.0.
   2. Cost of collection and disposal will double within 10 years (linear increase).
   3. Legislation will be passed within this period that will regulate, on a national level, the use and production of beverage containers.
   4. The GNP (in real dollars) is presently at 0.9 of its maximum, which it will reach in 5 years (at a linear increase), at which time it will decrease linearly at the same rate.
   5. The conditions underlying the historical trend increase will persist during the next 10 years.

Some of these assumptions could have resulted from UEJ methods (see Chapter 4), while others could have been obtained from other forecasts (e.g., government or research institute sources). It is implicitly assumed in this list that, only three factors will affect the generation rate in addition to trend:

(1) collection/disposal cost changes, (2) the enactment of beverage container legislation, and (3) real GNP changes.

We now estimate our future confidence in these assumptions. Using the approach previously described in Section 5.3, we suppose that our trend assumption is based on a hypothetical sample of N = 24, the cost of collection/disposal on N = 36, and GNP changes on N = 48. The future confidence is found from the appropriate curves in Figure 5-2 and entered into Table 5-2. This is done only for trend-type events. For dichotomous events (i.e., those that either occur or do not occur), such as the passage of beverage container legislation, the probability of enactment for any particular year is entered into Table 5-2 directly.

The next step in the algorithm is to determine the impact factors, c_j, for time = 0. These are determined by UEJ techniques, but at best these techniques can only produce relative estimates of the factors. We must first estimate the absolute value of at least one factor.

This is done in the following manner.

1. Assume the forecasted variable is at half its maximum value. This is f(0). (Since the forecast determined by Equations 5-5 and 5-6 follows a logistic curve, at half the forecast value the function is at its point of inflection, i.e., a neutral point with regard to growth and decay of the forecasted value.)
2. If the other variable is not dichotomous, the effect on the forecasted variable is for one time period because we assume a one time period change in the other variable. If the other variable is dichotomous (i.e., an "event"), the effect on the forecasted variable is that which occurs when we assume the event occurs.

Table 5-2. Forecasting Example of Section 5.6, Assumption Confidences

	Confidence in Assumptions (%)			
Year	A-1[a]	A-2[b]	A-3[c]	A-4[d]
0	99	99	0	99
1	97	98	0	99
2	93	97	0	98
3	88	95	0	98
4	83	93	0	96
5	77	91	0	95
6	70	88	0	94
7	64	85	10	92
8	59	82	30	90
9	55	78	60	88
10	51	74	0	86

[a]A-1 = underlying reasons that trend will continue (N = 24).
[b]A-2 = cost of collection/disposal (N = 36).
[c]A-3 = beverage container legislation.
[d]A-4 = GNP changes (N = 48).

3. Assess the effect on the forecasted variable of a change in the other variable, i.e., what will be the value of the forecasted variable now? This is $f(0 + 1)$.
4. Use Equation 5.5 to calculate $p(0)$, i.e.,

$$f(0 + 1) = f(0)^{p(0)}$$

5. Use the following modified form of Equation 5-7 to solve for c_h, i.e.,

$$p(0) = 1 + k\,|c_j|$$

It will be noted that the modification is simply to let the value of all other variables equal zero, and the value of x_j equal one.
6. The prior step supplies $|c_j|$. The sign of c_j is negative if x_j decreased the value of the forecast in step 3, positive if it increased its value.

As an example, suppose we select as the other variable the effect of collection/disposal on waste quantity generation. The steps are as follows, assuming a 10-year forecast period:

1. $f(0) = 0.5$.
2. The cost variable is not dichotomous; therefore, the effect to assess is for a one-year period. Since our second assumption was that cost would double (linear increase) within 10 years, the cost variable goes from 0.5 (scaled value on a 0–1 range) to 0.55 during the first time period.
3. We now suppose that this change in the cost variable from 0.5 to 0.55 produces a 10% decrease in waste generation, i.e., f goes from 0.5 at time 0 to 0.45 at time 1.
4. $0.45 = f(0 + 1) = f(0)^{p(0)} = (0.5)^{p(0)}$ and $p(0) = 1.1521$.
5. $1.1521 = p(0) = 1 + k|c_2|$. Since $k = 0.1$, $|c_2| = 1.5$.
6. Since the cost variable decreased the quantity of waste generated, $c_2 = -1.5$.

By this technique or by UEJ techniques using the value of c_2 as a comparison, the remaining impact factors are determined. With the exception of GNP, the impact factors do not change because the same conditions for the variables exist over the 10-year period. At year 6, however, the impact factor for GNP changes from 0.7 to −0.7 because our scenario has postulated a decrease symmetrical to the increase of the first 5 years. These values are entered in Table 5-3, as are the values from Table 5-2, which are taken as the estimates of x_j.

Table 5-3 is the computation form for formulas 5-5 and 5-7. Since the generation rate at time t = zero is 3.5 lb/person/day, f(0) becomes 0.7 when scaled in the 0–1 range. Repeated applications of Equation 5-5 completes the p(t) column, and use of Equation 5-7 completes the forecast column. To forestall any confusion about the symbolism of the last two columns, the p(t) refers to conditions at the beginning of time t, and the f(t) refers to the forecast at the end of time t. Thus, f(t) really is f(t + 1) in Equation 5-5, e.g.,

$$0.693 = (0.700)^{1.0261}$$

From Table 5-3 we see that the forecast at the end of year t = 10 is 0.376, which is equivalent to 0.188 lb/person/day in unscaled units. The complete

Table 5-3. Forecasting Example Computation Form

Time t	Trend x_1	Trend c_1	Cost x_2	Cost c_2	Legislation x_3	Legislation c_3	GNP x_4	GNP c_4	p(t)	f(t)
0	–	0.5	–	−1.5	–	−2.0	–	0.7	–	0.700
1	0.97	0.5	0.98	−1.5	0	−2.0	0.99	0.7	1.0261	0.693
2	0.93	0.5	0.97	−1.5	0	−2.0	0.98	0.7	1.0273	0.686
3	0.88	0.5	0.95	−1.5	0	−2.0	0.98	0.7	1.0269	0.679
4	0.83	0.5	0.93	−1.5	0	−2.0	0.96	0.7	1.0278	0.671
5	0.77	0.5	0.91	−1.5	0	−2.0	0.95	0.7	1.0285	0.663
6	0.70	0.5	0.88	−1.5	0	−2.0	0.94	−0.7	1.1573	0.621
7	0.64	0.5	0.85	−1.5	0.10	−2.0	0.92	−0.7	1.1764	0.571
8	0.59	0.5	0.82	−1.5	0.30	−2.0	0.90	−0.7	1.2089	0.508
9	0.55	0.5	0.78	−1.5	0.60	−2.0	0.88	−0.7	1.2638	0.425
10	0.51	0.5	0.74	−1.5	0	−2.0	0.86	−0.7	1.1421	0.376

10-year forecast is plotted in Figure 5-4 as the solid line. There is a linear decline until year t = 6, when the effect of GNP and the passage of beverage container legislation begins to be felt.

The effect of changes in assumptions can easily be made. Table 5-4 shows a recalculation of the problem where the only change is to shift the passage of legislation by four years. The initial values, confidences, probabilities and impact factors remain the same. The new forecast is shown as the dotted line in Figure 5-5. Although the end result is approximately the same, the decline in waste generation is experienced earlier.

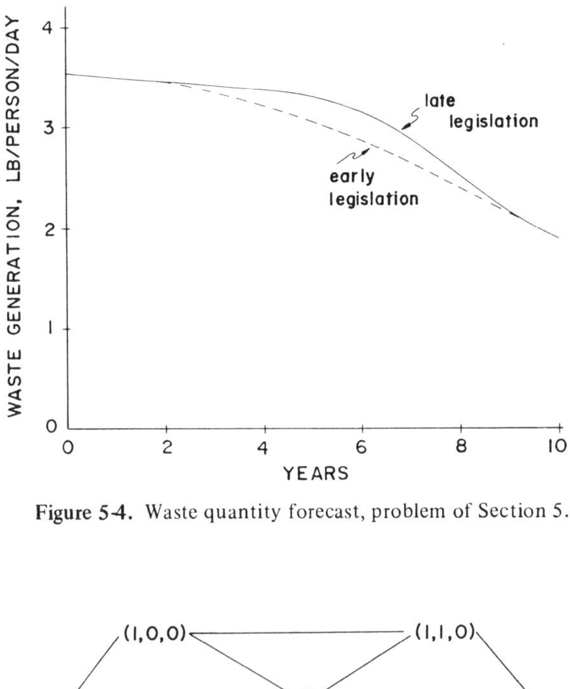

Figure 5-4. Waste quantity forecast, problem of Section 5.6.

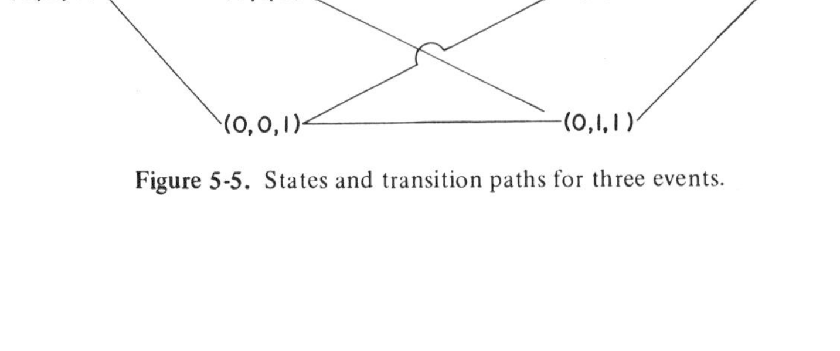

Figure 5-5. States and transition paths for three events.

Table 5-4. Forecasting Example Computation Form, Early Legislation Modification

Time t	Trend x_1	c_1	Cost x_2	c_2	Legislation x_3	c_3	GNP x_4	c_4	p(t)	f(t)
0	–	0.5	–	−1.5	–	−2.0	–	0.7	–	0.700
1	0.97	0.5	0.98	−1.5	0	−2.0	0.99	0.7	1.0261	0.693
2	0.93	0.5	0.97	−1.5	0	−2.0	0.98	0.7	1.0273	0.686
3	0.88	0.5	0.95	−1.5	10	−2.0	0.98	0.7	1.0448	0.675
4	0.83	0.5	0.93	−1.5	30	−2.0	0.96	0.7	1.0819	0.654
5	0.77	0.5	0.91	−1.5	60	−2.0	0.95	0.7	1.1371	0.617
6	0.70	0.5	0.88	−1.5	0	−2.0	0.94	−0.7	1.1573	0.572
7	0.64	0.5	0.85	−1.5	0	−2.0	0.92	−0.7	1.1549	0.525
8	0.59	0.5	0.82	−1.5	0	−2.0	0.90	−0.7	1.1520	0.476
9	0.55	0.5	0.78	−1.5	0	−2.0	0.88	−0.7	1.1471	0.427
10	0.51	0.5	0.74	−1.5	0	−2.0	0.86	−0.7	1.1421	0.378

5.7 FORECASTING EVENTS

The preceding section deals with the forecasting of quantitative variables measurable on a continuous scale. As was seen, such forecasts describe the (continuous) path of a response variable in time, based on the interrelationships (cross-impacts) with a set of other (predictor) variables and on the starting conditions of all variables. This section deals with nonrecurrent events, which may or may not occur in some future time interval under consideration. Nonrecurrent events occur only once if they do occur, but any recurrent event may be restated as a set of nonrecurrent events. (We can specify an exact number of occurrences; we can specify that the event will occur at least once, etc.). We can, of course, elicit a subjective probability that a particular event will take place within a stated time period, but the estimate would depend on a (usually) large number of other events that might take place. Under such circumstances it might be extremely difficult to determine which is a "response" event and which is a "predictor" event, to use an analogy with quantitative variables.

Consider, for example, the nonrecurring events E_1, E_2 and E_3. There are $2^N = 2^3 = 8$ distinct outcomes possible and $N2^{N-1} = 3(2)^{3-1} = 12$ transition paths (hence, probabilities) to these outcomes. This is shown in Figure 5-5, where the 3-tuples refer to events E_1, E_2 and E_3, respectively, with a zero indicating the event not occurring and a one indicating the event occurring. The number of transition paths increases rapidly with N. At $N = 5$, there are 80; at $N = 7$, 448; at $N = 9$, 2304; at $N = 10$, 5120. For large N, it is clearly not practical to estimate all of the transition probabilities. Even if it were possible to fashion a decision tree (see Chapter III) consisting of mutually exclusive outcomes, there still might be too many such outcomes for practical feasibility. For three events there are $2^N = 2^3 = 8$ possible outcomes as previously noted; for 10 events, however, there are 512 outcomes. Furthermore, in practice it may be extremely difficult to estimate the decision tree probabilities when the events involve economic, political and sociological aspects, as they do in many resource recovery decisions.

One approach to this problem is a version of cross-impact analysis suggested by Turoff [10]. It involves the following steps:

1. Specify a set of nonrecurring, interrelated events.
2. For each event, estimate the probability of that event occurring in the interval of time from now to some point in the future.
3. Perturb the event set by assuming a certainty that one of the events will occur, and reestablish the probabilities in step 2. Do this for all events in the event set.
4. Calculate the cross-impacts, based on steps 2 and 3.
5. By a cascading perturbation approach, forecast the occurrences or nonoccurrences of the events.

As seen from steps 2 and 3, this approach involves N^2 inputs for N events. For example, for 5 events, only 25 probabilities need be estimated; for 10 events, only 100.

It should be noted that the "conditional" probabilities asked for in step 3 are not the conditional probabilities of formal probability theory. Given two events, A and B, probability theory [11] states the following:

$$P(A|B) = P(AB)/P(B) \tag{5-9}$$

Suppose, for example, that event A is that "Congress will pass national beverage container legislation by 1985," and that event B is, "The cost of litter cleaning nationwide by 1985 will equal or exceed $1 billion." Assuming the probability of A is p_1, and of B is p_2, then

$$P(A) = p_1$$

$$P(B) = p_2$$

If it is certain that Congress will pass the legislation by 1985, then the probability that event B will occur is either unchanged or less than the original estimate (the law might be passed soon enough to affect litter cleanup costs), i.e.,

$$P(B|A) = p_3 \leqslant p_2$$

If it is certain that litter cleanup will reach or pass the $1 billion mark by 1985, then probably we would increase our estimate of A by some amount, say e:

$$P(A|B) - p_1 + e$$

where $e > 0$ and $P_1 + e \leqslant 1$. P(A|B) and P(B|A) were standard conditional probabilities, so from Equation 5-9,

$$P(AB) = P(A|B)P(B) = P(B|A)P(A)$$

However,

$$P(A|B)P(B) = p_1p_2 + ep_2$$

$$P(B|A)P(A) = p_3p_1 \leqslant p_2p_1 < P(A|B)P(B)$$

Therefore, P(A|B) and P(B|A) are not the usual sort of conditional probabilities.

The basis of the Turoff method is to assume the simplest algebraic equation that satisfies the boundary conditions of the problem, i.e., that if an event is certain to occur or not to occur, no change in the other events can influence the outcome of the certain event. This equation turns out to be the logistic equation of statistics, i.e., the Fermi-Dirac distribution in physics. The exact mathematical approach is given in the next section.

5.8 MATHEMATICAL FORMULATION OF CROSS-IMPACT APPROACH

We define the following terms as follows

P_i = probability that ith event will occur.
R_{ij} = probability that ith event will occur, given that jth event will occur.
C_{ij} = relative causal weight (cross-impact factor) of jth event on ith event.
G_i = a measure (gamma factor) of the effects of other events (on the ith event) not specified by the simple cross-impact factors.

Although P_i and R_{ij} must be positive, a negative C_{ij} or G_i indicates an inhibiting effect.

As an example, let us consider the following five events for the time interval now to 1985:

- E_1: National beverage container legislation will be passed.
- E_2: Litter disposal costs will reach $1 billion.
- E_3: Industry will start a massive recycling program for beverage containers.
- E_4: Environmental groups will start a strong, concerted lobby for national beverage container legislation.
- E_5: The annual inflation rate will exceed 10% for at least two consecutive years.

Let us assume that the P_i and the R_{ij} have been estimated for this problem (perhaps by UEJ techniques) and are shown in the probability table of Table 5-5. The probabilities of the events occurring are given in the diagonal of the table, i.e.,

$$P_1 = 0.50$$
$$P_2 = 0.75$$
$$P_3 = 0.20$$
$$P_4 = 0.25$$
$$P_5 = 0.50$$

The jth column gives the probability that the event in the ith row will occur, given that the jth event has occurred. We can also assume that the jth event has not occurred and estimate the change in original probabilities, but this complicates the computations. Therefore, for simplicity, only the "has occurred" statement is presented to the analyst here. For example, given that E_1 has occurred, the following are the probabilities of the other four events:

$$P\,(E_2/E_1 \text{ occurs}) = 0.95$$
$$P\,(E_3/E_1 \text{ occurs}) = 0.05$$
$$P\,(E_4/E_1 \text{ occurs}) = 0.40$$
$$P\,(E_5/E_1 \text{ occurs}) = 0.20$$

Table 5-5. Output for Example of Section 5.8

Probability Table

E	1	2	3	4	5
1	0.50	0.70	0.10	0.60	0.20
2	0.95	0.75	0.80	0.75	0.90
3	0.05	0.40	0.20	0.05	0.10
4	0.40	0.60	0.60	0.25	0.20
5	0.20	0.60	0.50	0.50	0.50

Cross-Impact Table

E	1	2	3	4	5
1	-0.74	3.29	-2.75	0.54	2.77
2	3.69	-1.92	0.36	0.0	2.20
3	-3.12	3.92	-1.44	-2.08	-1.62
4	1.39	6.02	1.88	-6.39	-0.58
5	2.77	1.62	0.0	0.0	0.17

Forecasting Certainty Sequence. The 1 Indicates Occurrence and the 0 Nonoccurrence

E	1	2	3	4	5
0	0.50	0.75	0.20	0.25	0.50
1	0.63	0.74	0.0	0.19	0.50
2	0.59	0.82	0.0	0.0	0.40
3	0.82	1.00	0.0	0.0	0.54
4	1.00	1.00	0.0	0.0	0.27
5	1.00	1.00	0.0	0.0	0.0

Forecasting Certainty Sequence. The 1 Indicates Occurrence and the 0 Nonoccurrence

E	1	2	3	4	5
0	0.50	0.75	0.80	0.25	0.50
1	0.10	0.80	1.00	0.60	0.50
2	0.0	0.39	1.00	0.50	0.81
3	0.0	0.65	1.00	0.06	1.00
4	0.0	0.65	1.00	0.0	1.00
5	0.0	1.00	1.00	0.0	1.00

For this problem we will assume the remainder of the probability table in Table 5-5. The cross-impact factors are calculated by

$$C_{ij} = [1/(1 - P_j)]\{\ln[R_{ij}/(1 - R_{ij})] - \ln[P_i/(1 - P_i)]\} \tag{5-10}$$

and the gamma factors are calculated by

$$G_i = \ln[P_i/(1 - P_i)] - \sum_{k \neq i} C_{ik}P_k \tag{5-11}$$

The derivation of these formulas is given in the Appendix, and the results are shown in the cross-impact table (Table 5-5). (The diagonal elements are the gamma factors.) If any probability is 0 or 1 (P_i or R_{ij}), these must be set to 0.01 and 0.99, respectively, before Equations 5-10 and 5-11 can be used.

To illustrate the use of Equation 5-10 and 5-11, we calculate the following:

$$C_{21} = [1/(1 - 0.50)]\{ln[0.95/(1 - 0.95)] - \ln[0.75/(1 - 0.75)]\} = 3.69$$

$$G_3 = ln[0.20/(1 - 0.20)]$$
$$- [(-3.12)(0.50) + (3.92)(0.75) + (-2.08)(0.25) + (-1.62)(0.50)] = -1.44$$

The next step is to forecast which events will and will not occur. This is accomplished by a cascading perturbation approach as follows:

1. Examine the overall probabilities (P_i) and determine which event(s) is closest to zero or one.
2. If the event is close to zero, assume it will not occur; if it is close to one, assume it will occur.
3. Based on the new probabilities found in step 2, recalculate the P_i by the following modified version of Equation 5-11:

$$ln[P_i/(1 - P_i)] = G_i + \sum_{k \neq i} C_{ik} P_{k(new)} \qquad (5\text{-}12)$$

4. Return to step 1 and continue the perturbation until all events have or have not occurred.

The results of the cascading approach are shown in the first Forecasting Certainty Sequence of Table 5-5. Note that E_3 is closest to zero or one (it is closest to zero), so that it is set equal to zero in perturbation 1. To illustrate the application of Equation 5-12, we calculate P_1 for perturbation 1.

$$ln[P_i/(1 - P_i)] = -0.74 + (3.39)(0.75) + (-2.75)(0) + (0.54)(0.25) + (-2.77)(0.50)$$
$$= 0.5525$$

and

$$P_i = 0.63$$

As can be seen from the forecasting sequence, events E_1 and E_2 are forecasted to occur, while E_3, E_4 and E_5 are forecasted not to occur.

It is also possible to examine the sensitivities of the forecast to changes in the overall probabilities, P_i. In the second Forecasting Certainty Sequence of Table 5-5, for example, we have changed the initial estimate of E_3 from 0.20 to 0.80, followed by the perturbation method just outlined. This time, events E_2, E_3 and E_5 are forecasted to occur, while E_1 and E_4 are forecasted not to occur.

Finally, the technique lends itself well to the group approaches (for the probability estimates) described in Chapter 4.

5.9 COMPUTER PROGRAM

Because the calculations are somewhat tedious, the complete procedure has been programmed in FORTRAN IV, a listing of which is shown in Figure 5-6. The input data set for the output of Table 5-5 is shown in Figure 5-7. The data deck is coded as follows:

- Card 1–number of events, N, FORMAT (I2)
- Next N Cards–probability table, FORMAT (25F4.2)
- Following Cards–revision of basic probabilities, FORMAT (25F4.2)

Note that the last line of Fi⁻ure 5-7 contains the diagonal probabilities of the initial probability table, with the probability of E_3 changed from 0.20 to 0.80.

```
         INTEGER FTVS(25),J,J1,J2,JC
      REAL FORCT(26,25),CIT(25,25),HIGH,LOW,PROBT(25,25)
C     READ IN N WHICH IS NUMBER OF EVENTS
C     IF N>25, FORCT,FTVS,CIT,AND PROBT MUST BE INCREASED. FORCT
C     SHOULD BE DIMENSIONED AS FORCT(N+1,N)
C       FORMAT STATEMENTS WOULD ALSO NEED TO BE CHANGED
        READ(5,5)N
 5      FORMAT(12)
        NP1=N+1
C      READS IN PROBABILITY TABLE AND CHECKS TO MAKE SURE THAT THE
C      PROBABILITY OF ANY EVENT IS NOT 0 OR 1. IF THEY ARE IT
C      CHANGES THEM TO .01 OR .99 RESPECTIVELY.
        DO 11 J=1,N
 11       READ(5,12)(PROBT(J,I),T=1,N)
 12     FORMAT(25F4.2)
        DU I I=I,N
          DO 1 J=1,N
            IF(PROBT(I,J) .EQ. 1.0) PROBT(I,J)=.99
            IF(PROBT(I,J) .EQ. 0.0) PROBT(I,J)-.01
C      CALCULATE THE CROSS IMPACT TABLE.
            CIT(I,J)=(1./(1.-PROBT(J,J)))*(ALOG(PROBT(I,J)/(1.-PROBT(I,J)
       &             ))-ALOG(PROBT(I,I)/(1.-PROBT(I,I))))
 1          CONTINUE
C       CALCULATE GAMMAS
        DO 30 1=1,N
          CIT(I,I)=ALOG(PROBT(T,T)/(1.-PROBT(I,I)))
          DO 20 K=1,N
            IF (K .EQ.I) GOTO 20
            CIT(I,I)=CIT(I,I)-CIT(I,K)*PROBT(K,K)
 20         CONTINUE
 30       CONTINUE
C       OUTPUT PROBABILITY TABLE
        WRITE(6,16)
 16     FORMAT('1',I5,'PROBABILITY TABLE')
        WRITE(6,7)(I,I=1,N)
        DO 19 I=1,N
          WRITE(6,8)I,(PROBT(I,J),J=1,N)
 19       CONTINUE
C       SET FIRST ROW OF FORECAST TABLE TO OVERALL PROBABILITIES
        DO 2 I=1,N
```

Figure 5-6. FORTRAN IV program for cross-impact forecasting technique. (This figure continues on the following two pages.)

FIGURE 5-6 (contd.)

```
 2        FORCT(1,I)=PROBT(I,I)
C       OUTPUT CROSS IMPACT TABLE
        WRITE(6,6)
 6      FORMAT(////I5,'CROSS IMPACT TABLE')
        WRITE(6,7)(I,I=1,N)
 7      FORMAT(' E',7X,25(I2,6X))
        DO 9 I=1,N
        WRITE(6,8)I,(CIT(I,J),J=1,N)
 8      FORMAT(3X,I2,3X,25(F5.2,3X))
 9      CONTINUE
C   INITIALIZE FORECAST TABLE VARIABLE SET. THIS ARRAY KEEPS
C   TRACK OF WHICH EVENT PROBABILITIES HAVE BEEN SET TO 0 OR 1
C   AND IGNORES THESE WHILE FINDING THE LARGEST AND SMALLEST
C   PROBABILITIES IN THE CURRENT ROW OF THE FORECAST TABLE.
 39  DO 40 I=1,N
  40   FTVS(I)=I
     DO 100 I=2,N
C    FIND THE SMALLEST AND LARGEST PROBABILITY IN ROW.
       HIGH=FORCT((I-1),FTVS(1))
       LOW=HIGH
       J1=1
       J2=1
       M1=N-I+2
       DO 60 J=2,M1
         X=FORCT((I-1),FTVS(J))
         IF(HIGH.GT.X) GOTO 50
         HIGH=X
         J1=J
         GOTO 60
 50      IF(LOW.LT.X) GOTO 60
         LOW=X
         J2=J
 60      CONTINUE
C    IS LOW CLOSER TO 0 THAN HIGH TO 1?
       IF ((1-LOW).LT.HIGH) GOTO 70
C    SET PROBABILITY TO 0
       FORCT(I,FTVS(J2))=0.
       JC=J2
       GOTO 80
C    SET PROBABILITY TO 1
 70    FORCT(I,FTVS(J1))=1.
       JC=J1
C    CALCULATE REMAINING PROBABILITIES
 80    DO 140 L=1,M1
          IF(L.EQ.JC) GOTO 140
          PHIP=CIT(FTVS(L),FTVS(L))
          DO 130 M=1,N
            IF(M.FQ.FTVS(L)) GOTO 130
            IF (M.EQ.FTVS(JC)) GOTO 160
            DELTA=CIT(FTVS(L),M)*FORCT((I-1),M)
            GOTO 170
160         DELTA=CIT(FTVS(L),M)*FORCT(I,M)
170         PHIP=PHIP+DELTA
 130        CONTINUE
          FORCT(I,FTVS(L))=EXP(PHIP)/(1.+EXP(PHIP))
 140      CONTINUE
       I1=I+1
C    SET REMAINING COLUMNS OF EVENT TO 0 OR 1
       DO 150 J=I1,NP1
 150     FORCT(J,FTVS(JC))=FORCT(I,FTVS(JC))
```

FIGURE 5-6 (contd.)

```
      N1=N-1
C     DROP EVENT NUMBER OUT OF FTVS
      DO 90 J=JC,N1
 90     FTVS(J)=FTVS(J+1)
100   CONTINUE
C     TAKE CARE OF LAST ITERATION
      IF (FORCT(N,FTVS(1)).GE. 0.5) GOTO 110
      FORCT((NP1),FTVS(1))=0.
      GOTO 120
110   FORCT((NP1),FTVS(1))=1.
120   CONTINUE
      WRITE(6,200)
C     OUTPUT FORECAST TABLE
200   FORMAT(////I5,'FORECASTING CERTAINTY SEQUENCE. THE 1 INDICATES',
     &' OCCURRENCE AND THE 0 NON-OCCURRENCE')
      WRITE(6,7)(I,I=1,N)
      DO 230 I=1,NP1
        N1=I-1
        WRITE(6,220)N1,(FORCT(T,J),J=1,N)
220     FORMAT(3X,I2,4X,25(F4.2,4X))
230     CONTINUE
C     READ IN NEW OVERALL PROBABILITIES AND RERUN FORECAST
      READ(5,12,END=500)(FORCT(1,L),L=1,N)
      GOTO 39
500   STOP
      END
```

5				
0.50	0.70	0.10	0.80	0.20
0.95	0.75	0.80	0.75	0.90
0.05	0.40	0.20	0.05	0.10
0.40	0.60	0.60	0.25	0.20
0.20	0.60	0.50	0.50	0.50
0.50	0.75	0.80	0.25	0.50

Figure 5-7. Input data for example of Section 5.8.

5.10 SUMMARY

Forecasting is by no means an exact science; however, as pointed out by Martin [7], the alternatives are worse:

1. *No forecast*: facing the future blindfolded;
2. *Anything can happen*: the future is a complete gamble, so there is no point in attempting to anticipate it;
3. *The glorious past*: look to the past and ignore the future;

4. *Window-blind forecasting*: things are on a fixed track and the only direction is "up";
5. *Crisis action*: wait until the problem has arrived; and
6. *Genius forecasting*: find an Ayatollah and ask him for his intuitive forecast.

The rational forecast, on the other hand, is teachable and learnable: it can be described and explained, and, most important, can be reviewed by others. The forecasting models presented in this chapter have a systematic, logical basis with their assumptions clearly stated. Other, more complex models could have been devised, but the author is not persuaded that additional sophistication would improve forecast accuracy. The models provide both for existing trends and postulated future events. They lend themselves as easily to group forecasting, utilizing UEJ techniques, as they do to forecasting by an individual.

APPENDIX TO CHAPTER 5

A 5.1 DERIVATION OF CROSS-IMPACT FORMULAS

We assume that there exists a set of N equations expressing the probability of each event as a function of the other N − 1 probabilities:

$$P_i = P_i(P_1, P_2, \ldots, P_k, \ldots, P_n) \text{ for } k \neq i \qquad \text{(A-1)}$$

If we now introduce new information (e.g., that another event has occurred or not occurred), then the change in this probability, dP_i, is given as

$$dP_i = (\partial P_i/\partial B)\Delta B + \sum_{k \neq i} (\partial P_i/\partial P_k)\Delta P_k \qquad \text{(A-2)}$$

where B is a collective measure of the impact of those events not included in the specified set of N events. Equation A-2 is merely the difference form of Equation A-1.

The boundary condition that each equation must satisfy is that if an event is certain to occur (or not occur), no change in the other events can influence the outcome of the certain event. In short,

$$\text{if } P_i = 1, \text{ then } \Delta P_i = 0 \qquad \text{(A-3)}$$

The simplest algebraic relationship that can satisfy this condition is

$$\partial P_i/\partial X = P_i(1 - P_i)\partial F_i/\partial X \qquad \text{(A-4)}$$

where X is any of the variables of differentiation on the right side of Equation A-2 (i.e., the P_k or the B), and F is an arbitrary function. Substituting Equation A-4 into Equation A-2 we obtain

$$\Delta P_i = P_i(1 - P_i)\left[(\partial F_i/\partial B)\Delta B + \sum_{k \neq i} (\partial F_i/\partial P_k)\Delta P_k\right] \tag{A-5}$$

The next simplifying assumption we make is that the partial derivatives with respect to the P_k's are constants, i.e.,

$$\partial F_i/\partial P_i = C_{ik} \tag{A-6}$$

Solving Equation A-5 as a differential equation we obtain

$$P_i = \frac{1}{1 + \exp\left(-G_i - \sum_{k \neq i} C_{ik}P_k\right)} \tag{A-7}$$

where G_i is a function of the unknown variables, B, and also incorporates a constant of integration.

We verify that Equation A-7 is a solution to Equation A-5 by taking the total derivative:

$$dP_i/dX = [(-1)/(1 + E)^2](dE/dX) = [(-E)/(1 + E)^2]\, d\left(-G_i - \sum C_{ik}P_k\right)\Big/ dX \tag{A-8}$$

where E is the exponent in Equation A-7. From Equation A-7,

$$E = (1 - P_i)/P_i \tag{A-9}$$

and

$$1 + E = 1/P_i \tag{A.10}$$

Substituting Equation A-9 and A-10 into A-8, and completing the derivative, we have

$$dP_i dX = \{[(1 - P_i)/P_i]/P_i^2\}\left(G_i' + \sum C_{ik}\right) \tag{A-11}$$

and

$$dP_i = P_i(1 - P_i)\left[(\partial F_i/\partial B)dB + \sum_{k \neq i} (\partial F_i/\partial P_k)dP_k\right] \tag{A-12}$$

Since

$$R_{ij} = P_i \text{ for } P_j = 1 \tag{A-13}$$

then from Equation A-7,

$$\ln[R_{ij}/(1 - R_{ij})] = G_i + \sum_{k \neq i,j} C_{ij}P_k + C_{ij} \tag{A-14}$$

Subtracting Equation A-14 from A-7 we have

$$C_{ij} = [1/(1 - P_j)]\{\ln[R_{ij}/(1 - R_{ij})] - \ln[P_i/(1 - P_i]\} \tag{A-15}$$

If we have obtained values for all the C's, then

$$G_i = \ln[P_i/(1 - P_i)] - \sum_{k \neq i} C_{ik}P_k \tag{A-16}$$

REFERENCES

1. Jantsch, E. "Toward a Methodology for Systematic Forecasting," *Technol. Forecasting* 1:409-419 (1970).
2. Lanford, H. W. "A Penetration of the Technological Forecasting Jungle," *Technol. Forecasting Social Change* 4:207-225 (1972).
3. Smith, H., and N. R. Draper. "Fitting a Straight Line by Least Squares," in *Applied Regression Analysis* (New York: John Wiley & Sons, Inc., 1966), p. 24.
4. Wilmot, R. C. "A Contribution to Forecast Theory," *Technol. Forecasting Social Change* 13:43-58 (1979).
5. Fisher, R. A., and F. Yates. "Table III. Distribution of t," in *Statistical Tables* (New York: Hafner Publishing Co., Inc., 1949), p. 32.
6. Winkler, R. E. "The Assessment of Prior Distributions in Bayesian Analysis," *J. Am. Stat. Assoc.* 62(319):776-800 (1967).
7. Martino, J. P. "Forecasting by Analogy," in *Technological Forecasting for Decisionmaking* (New York: American Elsevier Publishing Co., Inc., 1972), pp. 65-101.
8. Kane, J. "A Primer for a New Cross–Impact Language–KSIM," *Technol. Forecasting Social Change* 4(2):129-142 (1972).
9. Kane, J., W. Thompson and I. Vertinsky. "Health Care Delivery: A Policy Simulator," *Socio-Econ. Planning Sci.* 6:283-293 (1972).
10. Turoff, M. "An Alternative Approach to Cross Impact Analysis," *Technol. Forecasting Social Change* 3:309-339 (1972).
11. Fraser, D. A. S. "Simple Probability Distributions," in *Statistics, An Introduction* (New York: John Wiley & Sons, Inc., 1958), p. 30.

INDEX